APPLICATIONS OF SPATIAL STATISTICS

Edited by **Ming-Chih Hung**

Applications of Spatial Statistics

http://dx.doi.org/10.5772/61666
Edited by Ming-Chih Hung

Contributors

Khalid Al-Ahmadi, Janusz Szmyt, Yi-Hwa Wu, Ghulam H. Bham, Uday Manepalli, Beatriz Larraz, Jose Manuel Pavia, Monica Navarrete, Barbara Goličnik Marušić, Damjan Marušić

CBS, Edition **2018**

Published by InTech

Janeza Trdine 9, 51000 Rijeka, Croatia

Publishing Process Manager Andrea Koric

Technical Editor SPi Global

Cover Designer InTech Design Team

Additional hard copies can be obtained from orders@intechopen.com

Applications of Spatial Statistics, Edited by Ming-Chih Hung
p. cm.
Print ISBN 978-953-51-2756-7
Online ISBN 978-953-51-2757-4

Contents

Preface

Since quantitative revolution in 1950s and 1960s, spatial statistics has become an important part of techniques for analyzing spatial data. During the past six decades, spatial statistics has been used in many fields in geography and/or disciplines concerning our living environments, such as demography, medical geography, transportation, landscape ecology, precision agriculture and many others.

GIScience, remote sensing, or statistics software makes performing statistical calculations or tests simple—by just several mouse clicks. Better yet, most census or socio-economic data can be downloaded from the Internet with relative ease. This certainly increases the applicability of spatial statistics on topics that had not been associated with spatial statistics before. With the processing part being simplified, one should pay more attention to why and how, that is, the justification on using spatial statistics and interpretation of results.

This book is a collection of studies on applying spatial statistics in subjects such as transportation, precision agriculture, demography and ecology. Different studies require different aspects of spatial statistics. I hope these examples can inspire readers on the use and interpretation of spatial statistics, as well as caution on applying spatial statistics to these fields. Equally important is the assessment. After things are said and done, we still have to ask one basic question: From all numbers and calculations, one gets good results, but are they reliable? Assessment is to ensure your results and interpretations are reliable, but not just one set of assessments. Statistics sometimes can be tricky, if some of the fundamental assumptions are not met, and such assumptions can easily be overlooked in any part of the study. When dealing with spatial data, most users are comfortable with visualization in forms of maps, draws, images, etc. After all, a picture is worth a thousand words. In addition to quantitative assessment, it is essential to use GIScience and visualization technologies for assessment and explore things that cannot be seen by quantitative assessment alone.

Many people deserve hearty thanks for bringing this book to reality. I wish to extend my appreciation to all of the authors and reviewers who contributed to this book, to my colleagues, and to scholars and professionals around the world who inspired me in the study of geography, GIScience, and spatial statistics. I also want to extend my sincere gratitude to Ms. Andrea Koric from InTech, who tirelessly and patiently guided me through the entire project and provided me with essential and helpful resources. Without her help and guidance, this book is not possible. Finally, I thank my parents for their support on my education and my wife and children for their continued support and encouragement.

Dr. Ming-Chih Hung

Geography/Geographic Information Science

Northwest Missouri State University

Maryville, MO, USA

Chapter 1

Application of Spatial Statistics in Transportation Engineering

Uday R.R. Manepalli and Ghulam H. Bham

Additional information is available at the end of the chapter

http://dx.doi.org/10.5772/65051

Abstract

"Everything is related to everything else, but near things are more related than distant things" is the first law of geography. It can be hypothesized that spatially, occurrence of a crash can exhibit similarities. To identify spatial patterns of crashes, this chapter presents spatial autocorrelation techniques such as Moran's I and the Getis-Ord G_i*statistics; spatial interpolation such as kriging; and nonparametric probability density function and kernel density (K). The aim of this chapter is to provide application of spatial statistics in transportation engineering specifically to identify crash concentrations and patterns of clusters in a study area.

Keywords: The Getis-Ord Gi* statistics, Kernel-Density function, kriging, Moran's I, spatial autocorrelation, highway safety, crash

1. Introduction

In this chapter, spatial data analysis and its application in the field of transportation engineering specifically for crash data analysis is presented. Analysis of spatial data extends the representation of geographic space from discrete sets of points, lines and polygonal features to mapping surfaces characterizing a continuous space. Statistics using spatial relationships for the data mapped investigates the similarities among them. The first law of geography states "Everything is related to everything else, but near things are more related than distant things" [1]. This principle has been used in various fields such as criminology, economics, transportation, etc. to identify relationships within a geographic space. In order to perform spatial data analysis, geographical locations and attributes of an object (point, line or polygon, area) are required. Spatial data analysis can answer questions such as how spatial data distributions can be compared, and how future distributions based on current spatial data can be forecasted. In the past, different statistical techniques have been used with spatial data and they can be broadly classified as:

Spatial autocorrelation (SA): The basic principle of SA is similar to the first law of geography. SA is defined as the correlation of a variable with itself in space. SA measures the strength of autocorrelation and the assumption of independence. A variable is said to be spatially autocorrelated if there are systematic patterns in its spatial distribution. SA is positive if nearby areas (regions) are alike. Negative autocorrelation applies to neighboring areas that are unlike, and SA is not exhibited by random patterns.

SA is measured using spatial autocorrelation indices. Some of the commonly used indices are Moran's *I* and Geary's *C*. These indices are often referred to as global indices. They measure overall degree of spatial autocorrelation in a data set. For specific disaggregated estimates, local indices are used. Some of the local indices are local Moran's I [2], local Geary's C [3], and the Getis-Ord G_i^* statistics [4, 5].

Spatial interpolation: It is defined as the process of using data for locations to predict ones that are not sampled. Inverse distance weighting and kriging [6] are commonly used in spatial interpolation techniques. The latter considers a spatial lag relationship that has both systematic and random components.

Spatial regression: Due to spatial autocorrelation, ordinary regression models cannot be used. To identify the underlying effects between the dependent variable and a spatial lag of itself, geographically weighted regression (GWR) [7, 8] is used.

Additional analysis techniques include nonparametric analysis such as kernel density estimation [9], as used in point pattern analysis to identify the first-order effects, i.e., measure the variation in mean value.

This chapter is organized as follows: first, the fundamental concepts for several spatial statistics measures are explained, and it is followed by case studies related to the fundamental concepts. The chapter ends with conclusions and recommendations.

2. Fundamental concepts

This section presents the concepts related to spatial autocorrelation, i.e., Moran's I and the Getis-Ord G_i^* statistics; spatial interpolation, i.e., kriging; and nonparametric analysis, i.e., kernel density estimation. They are presented to show their use in transportation safety.

2.1. Moran's I

It is one of the oldest indicators of SA [2]. SA compares the value of a variable in one location with its value at other locations. Similar to a correlation coefficient, SA varies between −1.0 and +1.0. A positive correlation indicates clustering (i.e., higher crash concentrations in highway safety), whereas negative correlation indicates dispersion or low crash concentration. Moran's I is expressed as

$$Moran's\, I = \frac{n\sum_i \sum_j w_{ij}\left(Y_i - \bar{Y}\right)\left(Y_j - \bar{Y}\right)}{\left(\sum_{i\neq j} w_{ij}\right)\sum_i \left(Y_i - \bar{Y}\right)} \tag{1}$$

The term w_{ij} represents a contiguity matrix. If location j is adjacent to location i, the interaction receives a weight of 1; otherwise, zero. The term w_{ij} compares the sum of the cross products of values at different locations weighted by the inverse of the distance between the locations.

The significance of Moran's I can be evaluated by a Z value as

$$Z(I) = \frac{I - E(I)}{S(I)} \tag{2}$$

where $E(I)$, the expected value of Moran's I, can be computed as

$$E(I) = \frac{-1}{n-1} \tag{3}$$

$S(I)$, the standard deviation, is computed as

$$S(I) = \sqrt{\frac{n^2(n-1)s_1 - n(n-1)s_2 - 2s_0^2}{(n+1)(n-1)s_0^2}} \tag{4}$$

where

$$s_0 = \sum_{i\neq j} w_{ij} \tag{5}$$

$$s_1 = \frac{1}{2}\sum_{i\neq j}\left(w_{ij} + w_{ji}\right)^2 \tag{6}$$

$$s_2 = \sum_k \left(\sum_j w_{jk} + \sum_i w_{ik}\right)^2 \tag{7}$$

In the foregoing formula, *i*, *j*, and *k* represent the location of crashes. At a level of 5%, values of Z greater than +1.96 and less than −1.96 indicate significant positive and negative SA, respectively.

2.2. The Getis-Ord G_i* statistics

G-statistics, developed by Getis and Ord, analyzes the evidence of spatial patterns and represents a global SA index [4, 5]. The G_i* (pronounced as G-i-star) statistics, however, is a local SA index. It is more suitable for discerning clusters of high or low concentration. A simple form of the G_i* statistics is [10]

$$G_i^* = \frac{\sum_{j=1}^{n} w_{ij} x_j}{\sum_{j=1}^{n} x_j} \tag{8}$$

where G_i* is the SA statistics of an event *i* over *n* events (e.g., crashes) [11]. The term x_j characterizes the magnitude of the variable x at event j over all n, and in highway safety, an index such as crash severity index (CSI) value determined at a particular location can be used. The G_i* statistics can be observed from the underlying distribution of the variable *x* [11]. The threshold distance (the proximity of one crash to another) can be set to zero to indicate that all features were considered neighbors of all other features.

Further, the standardized G_i* is essentially a Z value as well and can be associated with statistical significance

$$G_i^* = \frac{\sum_{j=1}^{n} w_{ij} x_{ij} - \bar{X} \sum_{j=1}^{n} w_{ij}}{S \sqrt{\frac{\left[n \sum_{j=1}^{n} w_{ij}^2 - \left(\sum_{j=1}^{n} w_{ij} \right)^2 \right]}{n-1}}} \tag{9}$$

where

$$S = \sqrt{\frac{\sum_{j=1}^{n} x_j^2}{n} - \left(\bar{X}\right)^2} \tag{10}$$

Positive and negative G_i* statistics values correspond to clusters of crashes with high- and low-value events, respectively. A G_i* statistics close to zero implies a random distribution of events.

2.3. Kriging

Kriging, a spatial prediction methodology based on spatial interpolation, was first developed by Matheron [12] based on the work of Krige [6] to predict ore reserves. Kriging has been applied widely in air quality analysis, geology, hydrology, ecology, etc. The major application of this technique is to predict values at unmeasured locations while assessing the errors of these predictions [13]. It relies on the notion that unobserved factors are autocorrelated over space, and the levels of autocorrelation decreases with distance. A trend estimate, $\mu(s)$, is determined which can be defined as [13]

$$Z_i(s) = \mu_i(s) + \varepsilon_i(s) \tag{11}$$

where $Z_i(s)$ is the variable of interest and s indicates the location of the site "i." It is composed of a deterministic trend $\mu_i(s)$ and a random error term ε_i (s). The random errors are autocorrelated over space. The expected value of Z(s) results in different types of kriging, namely simple, ordinary, universal, intrinsic kriging, and so on. However, universal kriging is preferred to other kriging methods as the trends depend on explanatory variables and (unknown) regression coefficients. The correlation between Z(s) and Z(s + h) does not depend on actual locations, but only distance "h" between the two sites. This is possible by assuming weak stationarity in all three cases. This indicates a constant variance of 2γ (h) for any s and h, where γ (h) can be expressed as

$$\gamma(h) = \frac{1}{2} var\left[Z(s+h) - Z(s)\right] \tag{12}$$

where $var[Z(s + h) - Z(s)]$ is the variance between s and s + h. When 2γ (h) is plotted versus distance, the plot is called a semivariogram. A semivariogram depicts the spatial autocorrelation of the measured sample points. One of the major steps is to select an appropriate semivariogram model that best fits the relationship between γ and h. There are three models that best explain the relationship, i.e., exponential, spherical, and Gaussian. In this chapter, only spherical model is presented, and the specifications are

$$\gamma(h) = \begin{cases} c_0 + c_1\left[1.5\frac{h}{a}0.5\left(\frac{h}{a}\right)^3\right] if\, 0 < h < a \\ c_0 + c_1\, if\, h > a \\ 0\, otherwise \end{cases} \tag{13}$$

The different models (spherical, exponential, and Gaussian) rely on parameters that describe their shape and level of spatial autocorrelation in the data. c_0 in the above equation is called the nugget effect and reflects discontinuity in the variogram origin as caused by factors such as sampling error and short-scale variability. The origin of the term nugget originates from gold deposits, as gold commonly occurs as nuggets of pure metal that are much smaller than the size of a sample. It can result in strong variability in the sample when physically close, and therefore discontinuity of the variogram at the origin can be observed [14].

The rate of variogram reflects the degree of dissimilarity of more distant samples. At large distances, a variogram can increase indefinitely if the variability of the phenomenon has no limit. However, if the variogram stabilizes at a value, called the sill, it indicates that beyond a certain distance Z(s) and Z(s+h) are uncorrelated [14]. This distance is called the range denoted by a. It determines the threshold distance at which γ(h) stabilizes [13]. $c_0 + c_1$ is the maximum γ (h) value, called sill, and c_1 is referred to as partial sill [15]. **Figure 1** illustrates a semivariogram.

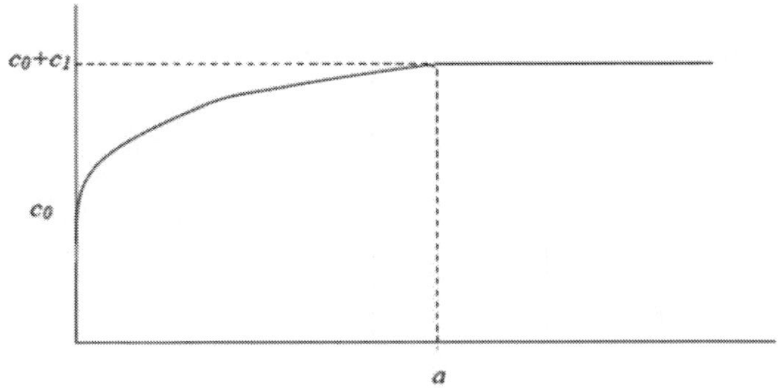

Figure 1. Illustration of a semivariogram.

2.4. Kernel density estimation

The kernel density method is a nonparametric method that uses a density estimation technique. It enables the observer to evaluate the local probability of an occurrence and degree of danger in a zone. For a given set of observations from an unknown probability density function, the kernel estimator can be defined as

$$\hat{f}(x) = \frac{1}{nh} \sum_{i=1}^{n} K\left(\frac{x - x_i}{h}\right) \quad (14)$$

where h is called the smoothing parameter or bandwidth, K is called the kernel, and $\hat{f}$ is the estimator of the probability density function f. Thus, the kernel estimator depends on bandwidth (h) and kernel density (K). For a given kernel, K, the kernel estimator critically depends on the choice of the smoothing parameter h. An appropriate choice of the smoothing parameter should be determined by the purpose of the estimate.

3. Case studies

The different case studies presented are related to the fields of crash data analysis, safety, and forecasting of traffic volume.

3.1. Spatial autocorrelation

A study was conducted to identify crash contributing factors on highway networks of Arkansas using a sample of crash data. In this study, spatial autocorrelation indices i.e., Moran 's I and Getis–Ord G_i^* statistics, and multinomial logistic regression were used. Autocorrelation was determined at different levels, and then multinomial logistic regression was used to identify crash-contributing factors in case a crash occurs. Based on the autocorrelation indices, the state's 75 counties were divided into zones. Further, to identify the crash contributing factors, a sample of data from the counties were compared to the statewide data.

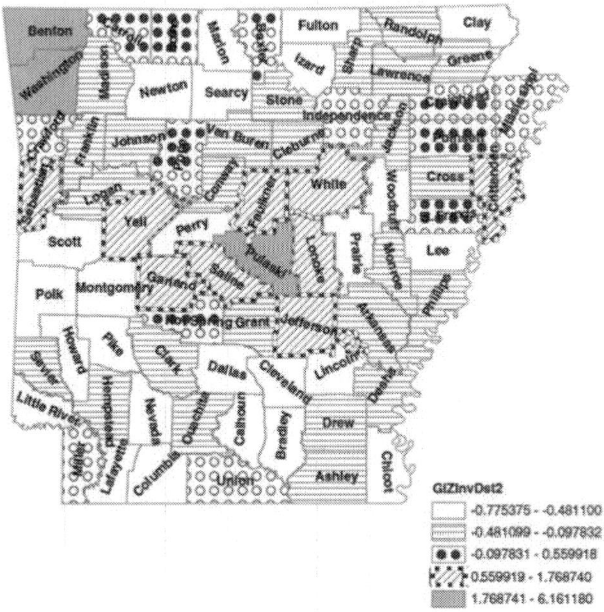

Figure 2. Counties categorized by G_i^* statistics [16].

Category	Number of counties	Counties with highest CSI[a]	CSI[b]	Total CSI[c]	CSI ratio[d]	Crash freq. ratio	G_i* statistics: range of Z values
(A)	(B)	(C)	(D)	(E)	(F)	(G)	(H)
First	3	Pulaski	137,627	276,755	.50	.51	1.7678741, 6.161180
Second	9	Garland	52,189	324,668	.16	.27	0.559918, 1.768740
Third	13	Craighead	28,676	298,379	.10	.17	−0.097831, 0.559918
Fourth	25	Madison, Cleburne, Logan	45,707	273,196	.17	.16	−0.481099, −0.097832
Fifth	25	Chicot, Montgomery, Polk, Perry, Little River, Clay, Colombia	53,477	133,861	.40	.57	−0.775375, −0.481100
Total	75	13[e]	317,676	1,306,859	.24	.34	–

Note: "–" not applicable.
[a]Satisfies the condition of minimum sample size of 2000 in terms of crash frequency.
[b]CSI computed for county/counties in Column C.
[c]CSI computed for counties in Column B.
[d]Ratio of CSI values in Columns D and E.
[e]Total number of counties in Column C.

Table 1. Results presented by category, highest CSI in each category, and ratios of crash data [16].

Crash data from 2004 to 2006 were used for the study. Crashes were categorized into five levels of crash injury severity from S1 to S5, where S1 indicated fatal injury; S2, major injury; S3, minor injury; S4, complain of pain; and S5, property damage only (PDO), based on the KABCO scale. Further, crash frequency (CF), i.e., the summation of crash count at various levels of crash injury severity, and crash severity index (CSI) [16] which combines various effects of different levels of crash injury severity into an index were determined. The first step of the analysis was to determine whether spatial autocorrelation exists. Moran's I was used which identified that SA exists for the crash data used. The crash injury severity levels showed significance at various levels.

G_i* was used to discern cluster structures of high or low concentration. Z-values were also computed and the categorization of counties based on the z-values of the G_i* statistic was determined. This categorization can be based on six different classification schemes: equal interval, defined interval, quartile, natural breaks, geometric interval, and standard deviation. The natural breaks scheme was best suited for the study [17]. In the natural breaks scheme, the classes are based on inherent categorizing in the data. The classes identify the break points that best groups similar values and maximizes the differences between these classes.

http://dx.doi.org/10.5772/65051

In the study, Jenks' algorithm was used to categorize the natural breaks [17]. Jenks' algorithm is commonly used to classify the data in a choropleth map, a type of thematic map that uses shading to represent classes of a feature associated with specific areas (e.g., a population density map). Jenks' algorithm generates a series of values that best represent the actual breaks in the data as opposed to some arbitrary classification scheme. Thus, it preserves the true clustering of data values. As a result, the algorithm creates "k" classes as the variance within categories is minimized. The state of Arkansas was categorized into five categories. **Figure 2** shows these categories, and **Table 1** presents the results by category, and shows the number of counties in each category. From each category, a county or a set of counties starting with the highest CSI was selected as a data sample. The highest CSI was used as the criterion because it provided the greatest variability in the crash data.

Figure 3 presents graphically the higher and lower Z values of G_i^* for the five categories. The G_i^* Z values indicate the clustering of the attributes in the study area. The first category had higher positive Z values compared to lower Z values, indicating that the value of CSI is not random for those counties. The trend from **Figure 3** indicates that the randomness increases over the categories. This trend is similar to the trend for identification of crash casual factors identified for each category, presented next.

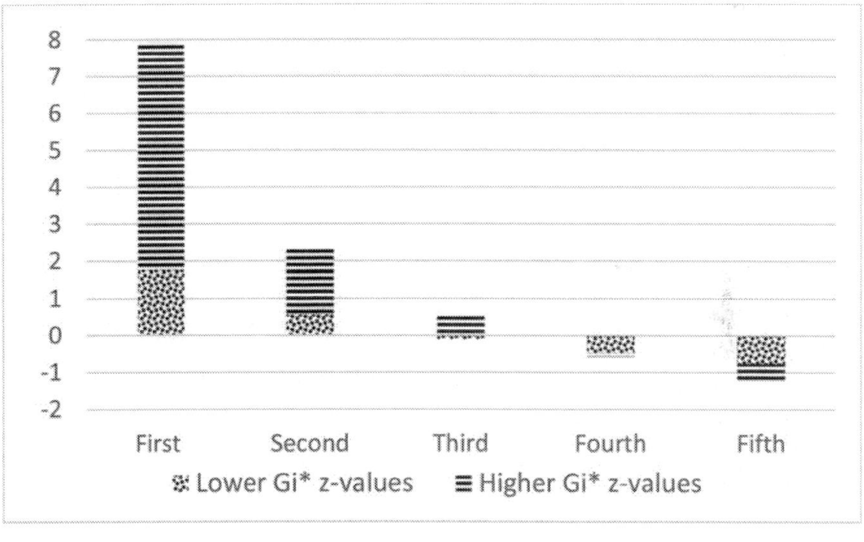

Figure 3. Comparison of G_i^* statistics values across five categories.

SA indices, however, do not explain why locations that indicate a cluster of crashes have a higher incidence of crashes compared with other locations; therefore, SA methods cannot identify crash causality factors [16]. Multinomial logistic regression (MLR) was used to identify the crash-contributing factors. The main reasons for choosing the MLR models were:

- Given that a crash has occurred, the factors that increase the chances of a fatal or a serious injury crash were considered and computed by using the odds ratio as a result of the MLR models.
- Factors that supplement the need for attainment of zero fatalities given that crashes occur because of other factors, including human factors, were identified.
- Factors for all levels of crash severity were identified, and common factors were selected as an alternate solution. However, this procedure is cumbersome when the desired results can be achieved in one model.
- A minimum sample size of 2000 is required to implement MLR models [18]. Therefore, with a decent sample size, these models can predict accurately. Details can be found elsewhere [18, 19].

Selected independent variables in the data were checked by using a variance inflation factor (VIF) to ensure that multicollinearity is not an issue. The variance inflation factor was found to be less than 10 for all of the variables; hence, multicollinearity was not observed. Variables selected for model development depended on the quality of the data. Only certain factors were retained for analysis since some factors had missing values. When more than 10% of the values were missing, that factor was not considered. For the factors presented in **Table 2**, no more than 1% of the values were missing. Mallows' Cp was used to retain the variables; a smaller value of Cp indicated a better model [19].

Abbreviations	Variables	Levels
ATM	Atmospheric conditions	Clear, rain
LGT	Light conditions	Dark, daylight
RSUR	Roadway surface	Dry, wet
RU	Roadway type	Rural, urban
RALI	Roadway alignment	Curve, straight
RPRO	Roadway profile	Grade, level
TOH	Roadway classification	Divided, undivided
TOC	Collision types	Angle, head-on, rear-end, sideswipe-same-direction (SSSD), single vehicle crashes (SVC), sideswipe–opposite direction (SWOD)
WK	Days of the week	Weekdays (M-F), weekends (Sat, Sun)
DUI	Driving under the influence	Yes, no
AADT	Annual average daily traffic	<20,000, 20,000–40,000, 40,000–60,000, 60,000–80,000, 80,000–100,000, 100,000–120,000

Table 2. List of independent variables [16].

Table 3 indicates that during darkness, fatal crashes were more likely to occur than PDO crashes, and the odds ratio increased by a factor of 1.28 if other variables remained constant. Similarly, the relative risk of fatal crashes was greater than the PDO crashes in rural areas and on curved roads.

Variables	Contributing factors	Estimate	Standard error	Chi-square value	p-Value	Odds ratio
Fatal vs property damage crashes						
LGT	Dark vs daylight	0.25	0.12	3.92	0.0476	1.28
RU	Rural vs urban	0.71	0.13	29.31	<.0001	2.04
RALI	Curve vs straight	0.34	0.13	6.25	0.0124	1.40
DUI	No vs yes	−1.17	0.13	86.71	<.0001	0.31
Major injuries vs property damage crashes						
Intercept		−2.49	0.18	185.59	<.0001	
RU	Rural vs urban	0.43	0.08	29.24	<.0001	1.54
RALI	Curve vs straight	0.29	0.08	13.05	0.0003	1.33
TOC	Angle vs SSSD	−0.39	0.17	5.36	0.0206	0.68
TOC	Head-on vs SSSD	1.86	0.23	64.43	<.0001	6.41
TOC	Rear-end vs SSSD	−0.58	0.15	15.34	<.0001	0.56
TOC	SVC vs SSSD	0.69	0.13	26.32	<.0001	2.00
TOC	SWOD vs SSSD	−1.50	0.25	36.62	<.0001	0.22
DUI	No vs yes	−0.77	0.08	91.27	<.0001	0.46

Table 3. Sample MLR results [16].

3.2. Kriging

Kriging models were used in a study to forecast Annual Average Daily Traffic (AADT) [13]. AADT data for 27,738 sites from 1999 to 2005 were used to forecast AADT values for 2006. The initial interpolation was made for 27,738 sites and later expanded throughout the network.

The study assumed that the AADT values would be similar to values at nearby sites. Network details were obtained based on the data provided by the Texas Department of Transportation. Two functional classes were identified Class 1 (interstate) and Class 2 (other principal arterials). Each site was then matched to attributes of the closest road section using functional class. Traffic counts on segments of the same class were spatially interpolated using kriging. For each functional class, a semivariogram was estimated. For Class 1 segments, the estimated range value, a, was 1.248; nugget value, c_0, was 2.33×10^7; and partial sill, c_1, was 1.62×10^7. For Class 2 segments, a equaled 0.158, c_0 9.86×10^8, and c_1 2.82 ×

10^9. It was found that Class 1 scatter was higher for a given distance compared to Class 2. The larger values of sill and nugget for Class 1 indicated spatial autocorrelation for AADT that is distance dependent and sensitive. Class 1 roads had many access points which might have led to fluctuations in AADT over space. For Class 2 roads, the flow changes appeared continuous over time.

The study concluded that more data helped improve the forecast, temporal dependence was stronger than spatial dependence, and kriging methods provided reliable results in uncounted/unsampled locations.

3.3. Kernel density estimation

A study examined the spatial patterns of pedestrian crashes to identify high crash zones. The study evaluated methods to rank these zones using a Geographic Information System (GIS) [20]. To identify these high crash zones, crash concentration maps were developed. The crash concentration maps based on density values used simple and kernel density methods. Five years of crash data (1998–2002) for Las Vegas metropolitan was used in the study. For this chapter, the scope is limited to identifying the crash concentrations using the Kernel density method.

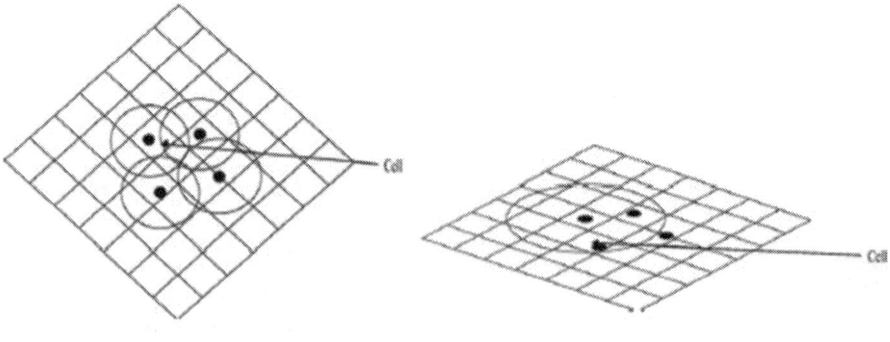

Figure 4. Illustration between kernel density (left) and simple density (right) methods [20].

The researchers identified the high crash zones using a three-step methodology: (1) geocode pedestrian crash data; (2) create crash concentration maps; and (3) identify zones, their shapes, and sizes. The geocoding of the crash data was performed using the "address match" feature. One of the major issues with point data, similar to crashes, is that when a map is plotted it may not present clusters of crash concentrations with more than just a few crashes. Developing maps with crash concentrations is therefore helpful.

Figure 4 illustrates the difference between simple density and kernel density methods, i.e., drawing a circular area of search around each crash to calculate the kernel values (K). The value of the surface is highest at the crash location and diminishes to zero at the radius of the circle. Thus, as a result, a smooth density surface is created.

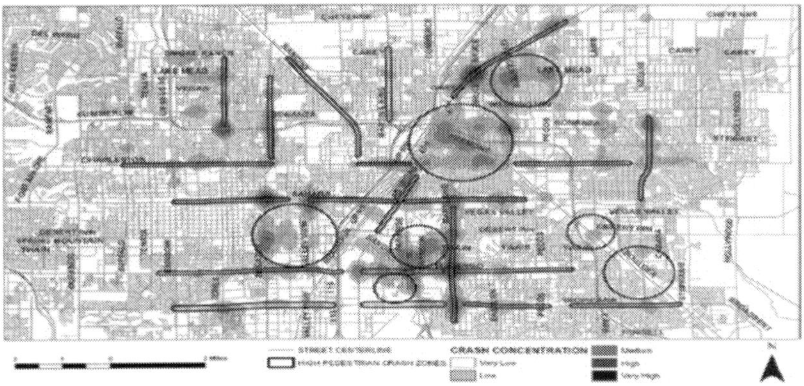

Figure 5. Las Vegas, pedestrian high crash zones [20].

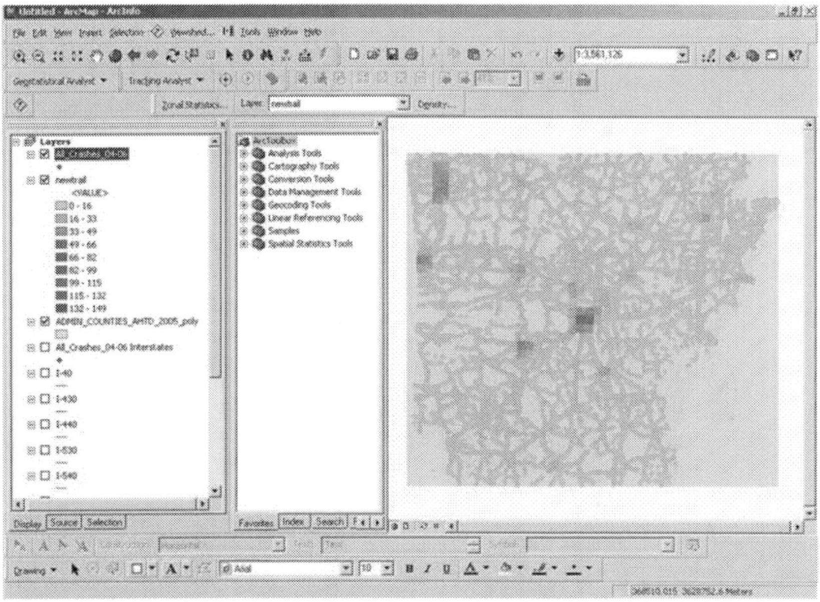

Figure 6. Identifying crash clusters using kernel density, application to Arkansas crash data.

Once the kernel density was identified, the zones of crash concentration were determined. These zones were either linear or circular. When dense clusters of crashes were observed along the route, then the zone identified was linear. When dense area was isolated at an intersection or was not linear in shape, then the zone identified was circular. When several linear zones were closely spaced and demographic, traffic, and geometric characteristics were similar, the researchers classified it as a circular zone. The study identified 29 pedestrian high crash zones, 22 linear zones, and 7 circular zones. **Figure 5** presents the 29 different crash zones.

The study concluded that the GIS-based methodology helps quantify the concentration of crashes and thus reduce the degree of subjectivity involved in identifying high crash zones. This approach is practical and easy to implement as most agencies collect crash, census, and traffic data sets in a GIS format.

In another study, undergraduate civil engineering students were exposed to the application of GIS in a mandatory course in transportation engineering [21]. The GIS tutorial was implemented in a laboratory environment developed as a self-guided activity supported by a web-based learning system. One of the tasks was to create a crash concentration map based on the data provided for a state highway network using the kernel density method. **Figure 6** presents a sample output from one of the students in the laboratory. The kernel density method is therefore easy to implement, and students in a laboratory when provided with a self-guided tutorial can implement it. The method when based in a GIS can also serve as a powerful tool to visualize crash clusters in a network.

4. Conclusions and recommendations

This chapter summarizes the fundamental concepts associated with spatial analysis of data in transportation engineering. Further, the application of these concepts is presented with interesting case studies from the literature specifically to improve highway safety and forecast of traffic volume for planning-level applications.

In various case studies presented in this chapter, a different spatial statistics model has been used. Depending on the type of problem, availability of data, expected outcomes, and ingenuity have led researchers to different techniques in spatial data analysis. These techniques help improve understanding of the phenomenon and thereby the solution to the problem. The future of spatial statistics lies in creative thinking and seeking solutions in more than one way. In terms of problem solving, solutions can be derived both objectively and subjectively. The more one experiments with the available techniques, the closer one can reach an ideal solution.

Author details

Uday R.R. Manepalli[1] and Ghulam H. Bham[2*]

*Address all correspondence to: ghbham@gmail.com

1 Agile Assets Inc., Austin, TX, USA

2 University of Alaska Anchorage, Anchorage, AK, USA

References

[1] Tobler W. A computer movie simulating urban growth in the Detroit region. Economic Geography, 1970; V. 46, 9: 234-240.

[2] Moran PAP. Notes on continuous stochastic phenomena. Biometrika, 1950; 37: 17-33.

[3] Geary R. The contiguity ratio and statistical mapping. The Incorporated Statistician, 1954; 5: 115-45.

[4] Getis A. Ord JK. The analysis of spatial association by use of distance statistics. Geographic Analysis, 1992; Vol. 24, No. 3: 189-206.

[5] Ord JK. Getis A. Local spatial autocorrelation statistics: distributional issues and an application. Geographic Analysis, 1995; Vol. 27, No. 4: 286-306.

[6] Krige DG. A Statistical Approach to Some Mine Valuations and Allied Problems at The Witwatersrand [Master's thesis]. Witwatersrand: University of Witwatersrand; 1951.

[7] Fotheringham A. S., Brunsdon C., Charlton M. E. Geographically Weighted Regression: The Analysis of Spatially Varying Relationships, Chichester, Wiley; 2002.

[8] Silverman B.W. Density Estimation for Statistics and Data Analysis, London, Chapman and Hall; 1985.

[9] Diggle, P. J., A kernel method for smoothing point process data. Journal of the Royal Statistical Society C, 1985; 34, 138-147.

[10] McGuigan DRD. The use of relationships between road accidents and traffic flow in "black-spot" identification. Traffic Engineering and Control, 1981; 448–453.

[11] Songchitruksa P, Zeng X. Getis-Ord Spatial Statistics for Identifying Hot Spots Using Incident Management Data. In Transportation Research Record: Journal of the Transportation Research Board, Transportation Research Board of the National Academies, Washington, D.C., 2010; 2165: 42-51.

[12] Matheron G. Principles of Geostatistics. Economic Geology, 1963; 58: 1246-1266.

[13] Wang X, Kockelman KM. Forecasting Network Data: Spatial Interpolation of Traffic Counts using Texas Data. In Transportation Research Record: Journal of the Transportation Research Board, Washington, D.C., 2009; 2105: 100-108.

[14] Chiles J-P., Delfiner P. Geostatistics: Modeling Spatial Uncertainty, 2nd ed., New Jersey, Wiley; 2012.

[15] Cressie N. Statistics for Spatial Data, New York, Wiley Interscience; 1993.

[16] Manepalli URR., Bham GH. Identification of Crash-Contributing Factors: Effects of Spatial Autocorrelation and Sample Data Size, Transportation Research Record: Journal

of the Transportation Research Board, Transportation Research Board of the National Academies, Washington, D.C., 2013; 2386: 179–188.

[17] Geography lecture notes [Internet]. Available from: http://go.owu.edu/~jbkrygie/krygier_html/geog_353/geog_353_lo/geog_353_lo07.html [Accessed: 2016-03-10].

[18] Ye F., Lord D. Investigation of Effects of Underreporting Crash Data on Three Commonly Used Traffic Crash Severity Models: Multinomial Logit, Ordered Probit, and Mixed Logit. In Transportation Research Record: Journal of the Transportation Research Board, Transportation Research Board of the National Academies, Washington, D.C., 2011; 2241: 51-58.

[19] Bham GH., Javvadi BS, Manepalli, URR. Multinomial logistic regression model for single-vehicle and multivehicle collisions on Urban U.S. Highways in Arkansas. Journal of Transportation Engineering, 2012; 138, No. 6: 786-797.

[20] Pulugurtha SS., Krishnakumar VK., Nambisan SS. New Methods to Identify and Rank High Pedestrian Crash Zones: An Illustration. Accident Analysis and Prevention, 2007; Vol. 39, No. 4: 800-811.

[21] Bham GH, Cernusca D, Luna R, Manepalli, URR. Longitudinal Evaluation of a GIS Laboratory in a Transportation Engineering Course, ASCE Journal of Professional Issues in Engineering Education and Practice, 2011; 137: 258-266.

Chapter 2

Comparison of Spatial Interpolation Techniques Using Visualization and Quantitative Assessment

Yi-Hwa (Eva) Wu and Ming-Chih Hung

Additional information is available at the end of the chapter

http://dx.doi.org/10.5772/65996

Abstract

Spatial interpolation has been widely and commonly used in many studies to create surface data based on a set of sampled points, such as soil properties, temperature, and precipitation. Currently, there are many commercial Geographic Information System (GIS) or statistics software offering spatial interpolation functions, such as inverse distance weighted (IDW), kriging, spline, and others. To date, there is no "rule of thumb" on the most appropriate spatial interpolation techniques for certain situations, though general suggestions have been published. Many studies rely on quantitative assessment to determine the performance of spatial interpolation techniques. Most quantitative assessment methods provide a numeric index for the overall performance of an interpolated surface. Although it is objective and convenient, there are many facts or trends not captured by quantitative assessments. This study used 2D visualization and 3D visualization to identify trends not evident in quantitative assessment. This study also presented a special case, a closed system in which all interpolated surfaces should sum up to 100%, to demonstrate the interaction between interpolated surfaces that were created separately and independently.

Keywords: spatial interpolation, quantitative assessment, 2D visualization, 3D visualization, performance

1. Introduction

Spatial interpolation is the process of using a set of point data to create surface data [1, 2]. A point data set has data values only for certain locations, such as field work locations, within the study area. Surface data divides the study area into cells, with a data value for each cell. With surface data, there is often a data value for every location inside the study area, whether it was sampled or not. Though a set of point data is more manageable in terms of labor, budget, and time; surface data are more useful and practical in many disciplines, such as precision agriculture, particularly with variable rate applications [3–9].

There are many spatial interpolation algorithms available in the literature, as well as in commercial GIS or statistics software [1, 10]. Each algorithm typically requires different parameters. Even with the same algorithm and same input data points, these different parameters can create different surfaces.

Evaluation of interpolated surfaces is difficult and often times overlooked. In most spatial interpolation studies, quantitative assessment was the only method used to evaluate the resultant surfaces. Most quantitative methods provide a numeric index for overall performance. Such a numeric index is easy to understand and convenient [10–13]. However, interpolated surfaces cannot be described by one numeric index, as many characteristics cannot be observed or evaluated by quantitative assessments. To date, there is no "rule of thumb" on which spatial interpolation techniques are most appropriate for certain situations [14].

The purpose of this chapter is to demonstrate a comprehensive approach to evaluate spatial interpolation, including: common quantitative assessment, 2D visualization, and 3D visualization. This chapter also presents a special case, a closed system consisting of three variables. Spatial interpolation techniques were applied to the three variables separately and independently to create surfaces. 2D visualization and 3D visualization then were used to evaluate whether the interpolated surfaces met the requirements for a closed system. This chapter is organized as follows: Section 2—study area and data, Section 3—spatial interpolation methods, Section 4—quantitative assessments, Section 5—2D and 3D visualization, Section 6—special case of a closed system, and Section 7—conclusions.

2. Study area and data

The study area is a 12.15 ha field located at the northwest Missouri State University R.T. Wright Farm near Maryville, MO, USA (**Figure 1**). This field was managed under a corn-soybean rotation [14]. Soils within the field were mapped as mollisols. Soil samples were collected in January 2006 using five soils sampling schemes outlined in a previous study [15]: 0.11 ha grid with 110 samples, 0.98 ha grid with 12 samples, 3.04 ha grid with four samples, topography-based composite with three samples, and whole-field composite with only one sample. The soil pH value of the 110 sample points from the 0.11 ha grid was the input data for spatial interpolation in this study. Point sampling was used to collect grid-based samples; five 1.27 cm diameter soil cores to a 15.24 cm depth were randomly collected from a 1.0 m^2 area around the predetermined cell sample point. The five soil cores were composited to form the sample for each respective grid sample location. Soil pH was as determined using the standard laboratory method of the United States Department of Agriculture [16].

http://dx.doi.org/10.5772/65996

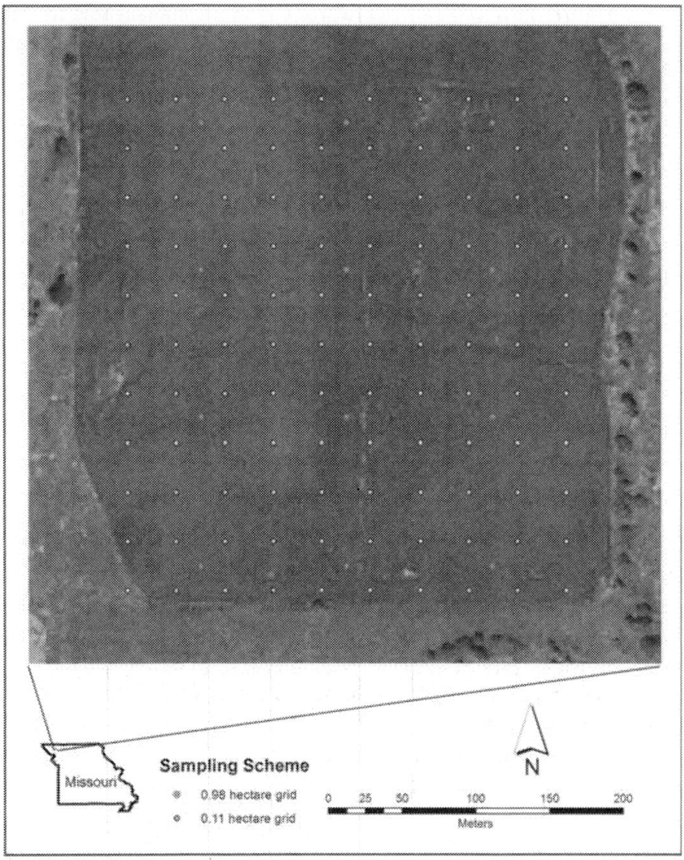

Figure 1. Study area: R. T. Wright University Farm in northwest Missouri, with NAIP (National Agricultural Imagery Program) 2006 CIR (color infrared) display.

3. Spatial interpolation methods

Spatial interpolation, or spatial prediction, is a process to estimate values of locations that were not surveyed based on a network of points with known values [1, 2, 10, 11]. In most cases, the input data is a network of points, while the output is a surface that divides the study area into small cells with a data value for each cell. There are two basic assumptions for spatial interpolation. First is spatial autocorrelation, which is best explained by Tobler's first law of geography "everything is related to everything else, but near things are more related than distant things" [17]. The second assumption is that values are smooth and continuous over space. Many spatial interpolation techniques were developed based on these two assumptions. Commercial GIS or statistical software provides several spatial interpolation functions, such as inverse distance weighted (IDW), kriging, spline, and others.

Although there are many options for spatial interpolation, to date, there is no "rule of thumb" on which technique is best under what certain circumstances. Even with the same technique and same input point data, different parameters may result in different surfaces. Potentially, a given set of points and a given spatial interpolation technique can generate many different surfaces [10, 14]. Therefore, it is important to evaluate and understand the accuracy and reliability of surface data generated from spatial interpolation. In this study, IDW, kriging, and spline will be used to demonstrate the process to evaluate and visualize spatial interpolation surfaces.

3.1. Inverse distance weighted

Inverse distance weighted is a deterministic estimation method where values at unmeasured points are determined by a linear combination of values at nearby measured points. Among available parameters, the power parameter can significantly affect the results. As the power parameter increases, IDW acts similarly to the nearest neighbor interpolation method in which the interpolated value is close to the value of the nearest measured value. The advantages of IDW are that it is simple, easy to understand, and efficient. Disadvantages are that it is sensitive to outliers and there is no indication of error [1].

Schloeder et al. [18] compared IDW, kriging, and spline spatial interpolation methods. They concluded that IDW and kriging performed similarly and that both are more accurate than the spline interpolation method. Mueller et al. [19] compared IDW and kriging on soil properties. Though individual performance differed greatly depending on the existence of spatial structure and sampling density, they concluded little difference between the overall performances between IDW and kriging. Kravchenko [20] conducted another study to compare IDW and kriging on soil properties. He reported that spatial structure significantly affected the accuracy of interpolation performance. He also reported that known variograms can greatly improve kriging performance, which may result in a better performance than IDW. Lu and Wong [21] developed a new form of IDW, which estimated data values at an unsampled location based on spatial pattern found in its neighborhood. As already reported in Refs. [19, 20], Lu and Wong [21] also found that variograms may greatly affect the performance of kriging. Their new form of IDW may perform better than kriging without variograms.

3.2. Kriging

Kriging is a stochastic method similar to IDW in that it also uses a linear combination of weights at known locations to estimate the data value of an unknown location. Variogram is an important input in kriging interpolation. It is a measure of spatial correlation between two points. With known variograms, weights can change according to the spatial arrangement of the samples. A major advantage of kriging is that, in addition to the estimated surface, kriging also provides a measure of error or uncertainty of the estimated surface. A disadvantage is that it requires substantially more computing time and more input from users, compared to IDW and spline [1].

Bekele et al. [22] compared several spatial interpolation methods, including kriging and IDW. They found that kriging generally performed better than IDW. However, they concluded that a regression-based autocorrelated error model was overall a more flexible method for interpolation. Laslett et al. [23] compared kriging and spline spatial interpolation methods and found that kriging produced better and more accurate surface than spline. Gotway et al. [24] compared kriging and IDW, and reported that kriging performed better than IDW and was relatively more stable because it was less dependent on spatial structure or soil sampling. Bishop and McBratney [25] conducted a study to explore the effect of having secondary data (such as color aerial photos) in the interpolation process. They reported an improved kriging performance.

3.3. Spline

Spline is a deterministic method to represent two-dimensional curves on three-dimensional surfaces. It can be imagined as fitting a flexible surface through a set of known points using a mathematical function. A major advantage of spline is that it can create fairly accurate and visually appealing surfaces based on only a few sample points. Disadvantages of spline are that the resultant surface may have different minimum and maximum values from the input data set, it is sensitive to outliers, and there is no indication of errors [1].

Laslett et al. [26] conducted an early study to evaluate and compare the performance of different spatial interpolation methods, including kriging, IDW, spline, and others. They reported though each method may perform better than others under certain situations, overall spline and kriging performed relatively better than IDW. Voltz and Webster [27] compared kriging and spline on soil properties, and concluded that kriging performed overall better than spline. Robinson and Metternicht [28] compared spline, kriging, and IDW interpolations methods on soil properties. They reported that no single method was suitable for all situations. Simpson and Wu [29] compared IDW, kriging, and spline on interpolating lake depth, and reported that spline produced the most accurate results with less than the ideal amount of sampled points.

4. Quantitative assessment

Based on a previous study [14], six interpolated surfaces were chosen for demonstration purposes. They are IDW (parameters: power 2, 10 neighbors), spline (parameters: tension, 10 neighbors), kriging (parameters: circular, 10 neighbors), IDW (parameters: power 4, 20 neighbors), spline (parameters: thin plate, 20 neighbors), and kriging (parameters: exponential, 20 neighbors). Each surface was evaluated by cross validation (Jackkniffing) by the 110 points from the 0.11 ha grid [10]. This validation process will go through iterations till all points were processed and validated. In each iteration, one sample point with known data value was discarded, and the remaining sample points were used to predict the value at the location of the discarded point. The known data values were compared to their counterpart predicted values and a measure of prediction accuracy was calculated.

Four error measures were used as accuracy index [14]. They are (1) mean absolute error (MAE), see Eq. (1) [12, 30]; (2) root mean square errors (RMSE), see Eq. (2) [12]; (3) systematic root mean square errors (RMSEs), see Eq. (3) [31]; and (4) unsystematic root mean square errors (RMSEu), see Eq. (4) [31]. Readings from the accuracy index, the lower values mean less errors, and therefore, higher accuracies and better performances.

$$\frac{\sum_{i=1}^{n} |Pi - Si|}{n} \tag{1}$$

where n is the sample size, Pi is the predicted value at point i, and Si is the sampled value at point i.

$$\sqrt{\frac{\sum_{i=1}^{n} (Pi - Si)^2}{n}} \tag{2}$$

$$\sqrt{\frac{\sum_{i=1}^{n} (\hat{P}i - Si)^2}{n}} \tag{3}$$

where $\hat{P}i$ is the estimated value at point i, by the best-fit regression function specific to each interpolation surface.

$$\sqrt{\frac{\sum_{i=1}^{n} (Pi - \hat{P}i)^2}{n}} \tag{4}$$

Table 1 summarizes these four error measures for these six interpolated surfaces. At first glance, they are quite compatible with each, meaning a similar performance. With closer examinations, one may notice that spline (parameter: thin plate, 20 neighbors) seems to have higher error measures, meaning more errors, and therefore worse performance. This particular interpolation has 0.3481 in MAE measure, while other surfaces are between 0.2925

	MAE	RMSE	RMSEs	RMSEu
IDW, power 2, N 10	0.2930	0.3671	0.3164	0.1712
Spline, tension, N 10	0.2957	0.3702	0.3279	0.1813
Kriging, circular, N 10	0.2926	0.3669	0.3255	0.1669
IDW, power 4, N 20	0.2965	0.3702	0.3310	0.1815
Spline, thin plate, N 20	0.3481	0.4408	0.3508	0.3167
Kriging, exponential, N 20	0.2925	0.3661	0.3357	0.1540

N: neighbor parameter.

Table 1. Cross validation (Jackknifin g) by 110 sample points from 0.11 ha grid.

and 0.2965; 0.4408 in RMSE measure while others between 0.3661 and 0.3702; and 0.3167 in RMSEu measure while others between 0.1540 and 0.1815. Among these four error measures, spline (parameter: thin plate, 20 neighbors) interpolation has considerably higher values than the other surfaces in three measures. On the other hand, IDW and kriging seem to perform similarly with compatible error measures.

5. Visualization of spatial interpolation

5.1. 2D visualization

Figure 2 shows these six interpolated surfaces in a flat 2D visualization environment. With visual inspection, one may notice that among these three surfaces with 10 neighbors, kriging (parameter: circular, 10 neighbors) appears differently. One may describe it as smoother with less extreme values (because of less red colors and blue colors). On the other hand, IDW (parameter: power 2, 10 neighbors) and spline (parameter: tension, 10 neighbors) seem to appear similarly. The same observation can be made in the group of three surfaces with 20 neighbors. Kriging (parameter: exponential, 20 neighbors) appears smoother than other two surfaces. IDW (parameter: power 4, 20 neighbors) and spline (parameter: thin plate, 20 neighbors) seem to appear similarly. Comparison between the group of 10 neighbors and the group of 20 neighbors, one may observe another interesting trend that the group of 20 neighbors generally appears to have more extreme values, with more red colors and blue colors, than the group of 10 neighbors.

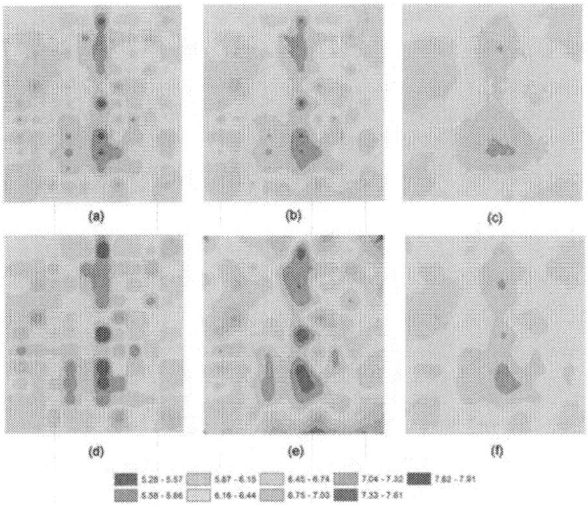

Figure 2. Six interpolated surfaces with their parameters. N: neighbor. (a) IDW, power 2, N 10; (b) spline, tension, N 10; (c) kriging, circular, N 10; (d) IDW, power 4, N 20; (e) spline, thin plate, N 20; (f) kriging, exponential, N 20.

Appearing smoother with less extreme values is not necessarily an indication of good performance or bad performance. It is just a characteristic of the overall trend of the interpolated surface, which was not revealed by quantitative assessment, such as four error measures shown earlier. An initial visual inspection of the interpolated surfaces already revealed a different observation from quantitative assessment. In quantitative assessment, it was observed that IDW and kriging performed similarly, and both are better than spline. With initial visual inspection, it was observed that IDW and spline performed similarly, while kriging performed differently, not necessarily in a better or worse way. Such difference warrants a further examination with visualization tools.

5.2. 3D visualization

Figure 3 shows the group of three surfaces with 10 neighbors in 3D visualization. One can confirm the trend observed in the 2D visualization that kriging (parameters: circular, 10 neighbors) appears smoothers than IDW (parameters: power 2, 10 neighbors) and spline (parameters: tension, 10 neighbors). This particular 3D visualization reveals even more trends that cannot be observed in quantitative assessment. In **Figure 3**, gray bars indicate locations of sample points, with bar height equaling data values. One may notice that kriging (parameters: circular, 10 neighbors) does not quite match the sampled data. Bars poke out (or appear above) the interpolated surface, indicating that the interpolated surface has data values less than actual sampled data. This is an indication of inexact interpolation [10], meaning the predicted data value at the sampled location is different from actual data value sampled at this same location. It implies that kriging (parameters: circular, 10 neighbors) underestimated data values, compared to actual data values. This phenomena (bar poking out of the surface) is less evident for spline (parameters: tension, 10 neighbors), and almost nonexistent for IDW (parameters: power 2, 10 neighbors). This implies that, in this study, kriging and spline are inexact interpolations, while IDW is an exact interpolation. There are parameters that can control exact or inexact interpolation in kriging or spline. Unfortunately, for most food producers, novice GIS users, or the general public, they are not familiar with exact or inexact interpolation. Chances are they do not know how to control the exact or inexact interpolation, and will end up like this study with some inexact interpolations, which is not revealed in quantitative assessment.

Figure 4 shows the group of three surfaces with 20 neighbors in 3D visualization. One can observe the same trend that kriging (parameters: exponential, 20 neighbors) appears smoothers than other two surfaces, with evident bars poking out of the surface. Comparing the group of 10 neighbors and the group of 20 neighbors, one may notice a difference in overall surface appearance. Taking IDW for example, IDW (parameters: power 2, 10 neighbors) has some pointy peaks, while IDW (parameters: power 4, 20 neighbors) appears duller. Same can be observed between spline (parameters: tension, 10 neighbors, pointy) and spline (parameters: thin plate, 20 neighbors, duller). One may also notice another abnormality on the south and north edges of spline (parameters: thin plate, 20 neighbors). There are some extreme peaks or villages among these two edges. This is also visible in the 2D visualization in **Figure 2(e)**, where some clusters of blue colors appear along the south edge and the north edge of the study area. Such clusters of blue colors are only visible in this particular interpolation.

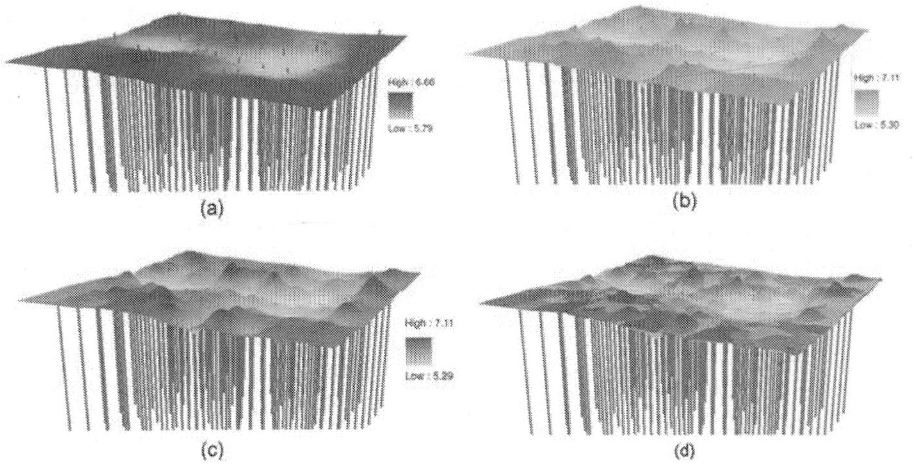

Figure 3. 3D visualization of three interpolations with 10 neighbor points. Each interpolation is displayed with a continuous tone, lighter colors for lower values, and stronger colors for higher values. View at the image from southwest. Soil sample data are displayed as gray bars, height of bars indicates data values. (a) Kriging, circular; (b) spline, tension; (c) IDW, power 2; (d) kriging, circular, spline, tension, and IDW, power 2 three interpolations.

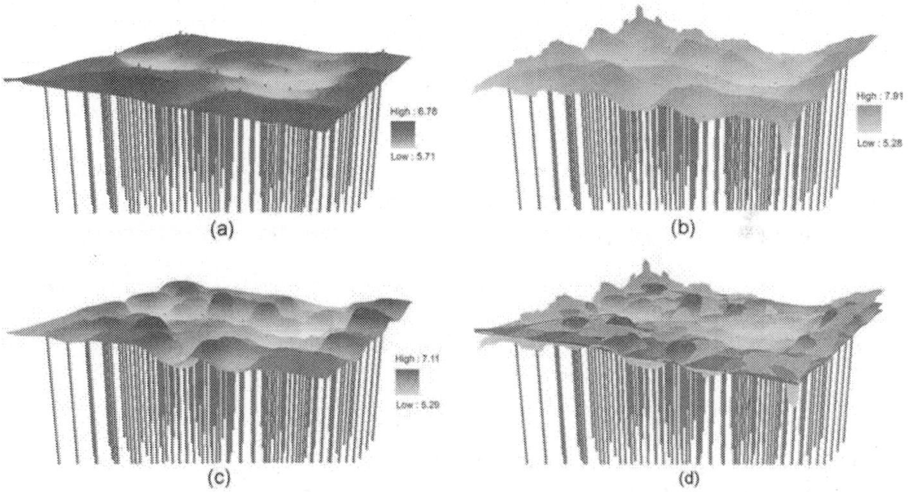

Figure 4. 3D visualization of three interpolations with 20 neighbor points. Each interpolation is displayed with a continuous tone, lighter colors for lower values, and stronger colors for higher values. View at the image from southwest. Soil sample data are displayed as gray bars, height of bars indicates data values. (a) Kriging, exponential; (b) spline, thin plate; (c) IDW, power 4; (d) kriging, exponential, spline, thin plate, and IDW, power 4 three interpolations.

	Min.	Max.	Mean	S.D.
IDW, power 2, N 10	5.29	7.11	6.30	0.24
Spline, tension, N 10	5.30	7.11	6.34	0.23
Kriging, circular, N 10	5.80	6.65	6.30	0.17
IDW, power 4, N 20	5.29	7.11	6.31	0.31
Spline, thin plate, N 20	5.29	7.91	6.34	0.35
Kriging, exponential, N 20	5.71	6.78	6.30	0.18
110 samples from 0.11 ha	5.29	7.11	6.27	0.38

N: neighbor points.

Table 2. Descriptive statistics for six interpolation results and the original sample set

Table 2 shows the descriptive statistics for these six interpolated surfaces, as well as the original sample data set (110 points from 0.11-ha grid). One may notice that only IDW surfaces have the exact minimum and maximum values as the original sample data. Overall, kriging has a smaller range (difference between minimum and maximum) than spline. Spline (parameters: thin plate, 20 neighbors) has the largest range, as observed in **Figures 2(e)** and **4(b)**.

In summary, different assessment methods reveal different characteristics of these interpolations. The quantitative assessment indicated that IDW and kriging performed similarly, and both better than spline. 2D visualization indicated that IDW and spline performed similarly, while kriging performed differently, not necessarily in a better or worse way. 3D visualization indicated that IDW is an exact interpolation, while kriging and spline are inexact interpolations. It was also revealed that kriging has the tendency to underestimate data values, compared to actual data values. Spline had the tendency to generate extreme data values along edges of the study area. Quantitative assessment is widely and commonly used in most spatial interpolation studies. Although 2D and 3D visualization tools do not provide quantitative indication of good or bad performance, they both revealed something quantitative assessment failed to report.

6. Interactions between spatial interpolations

So far, we have examined spatial interpolations on the individual surface level. As discussed earlier, it is difficult to determine which one performed better than others, based on one assessment method. Different assessment methods reveal different characteristics of interpolations. It is essential to understand these interpolated surfaces from all available assessment methods.

There are occasions where spatial interpolations were used to estimate a single variable in a larger project where multiple variables consist of a closed system. The V-I-S (vegetation-impervious surface-soil) model commonly used in modeling physical urban areas [32–34] is an example of such a closed system. In the V-I-S model, urban areas are represented by composition of vegetation, impervious surface, and soil. For example, industrial areas may be made of 50% impervious surface, 20% vegetation, and 30% soil, while low density residential areas may be made of 30% impervious surface, 60% vegetation, and 10% soil. The sum of V, I, and S percentage should be 100%, i.e., a closed system. When surveying V, I, and S percentage with field work, image processing, or photo interpretation, one can assure that surveyed data values sum up to 100%, meeting the closed system requirements. When doing the spatial interpolation to generate surfaces of V, I, and S percentages, special attention should be paid to the interactions between variables or surfaces.

6.1. Data and spatial interpolation in a closed system

A small experiment was conducted to demonstrate how individual spatial interpolation interacts with each other on a closed system. Fifteen points were visited and V, I, and S percentages were sampled in a grass field in Northwest Missouri State University in Maryville, MO, USA (see **Figure 5**). This field is grassy, with scattered trees, bushes, and pitches of

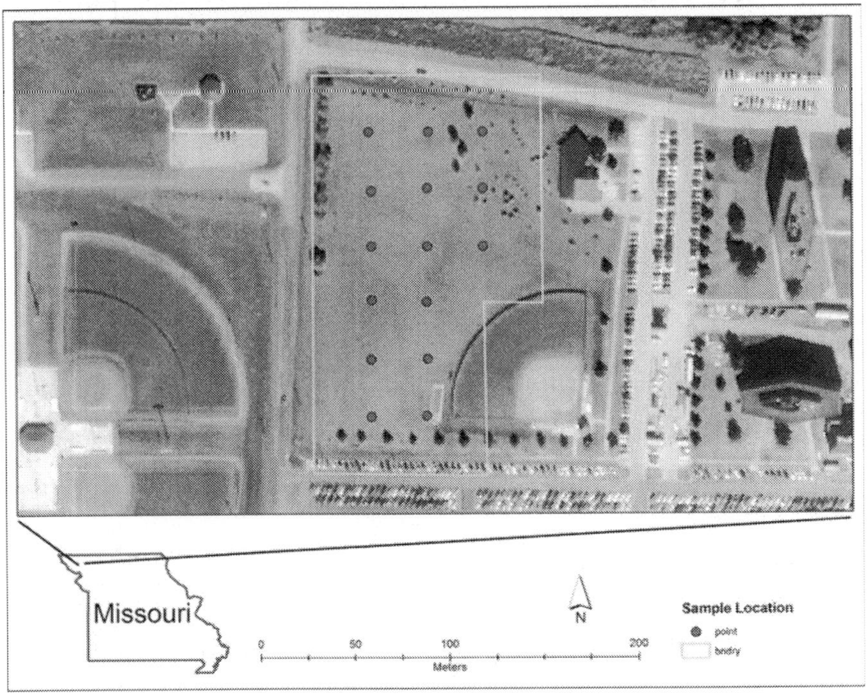

Figure 5. Study area: a grass field in Northwest Missouri State University, with 2003 IKONOS image (R, G, B/3, 2, 1) true color display.

soil. Impervious surface can only be found on the edges (roads and parking lots). Each point is 30 m away from its immediate four neighbors. At each point location, 100 samples were taken, with each sample verified as either vegetation, impervious surface, or soil. All 100 samples were then summed and converted to V, I, and S percentage for that point location. Most points have various amounts of vegetation and soil, with no impervious surface,

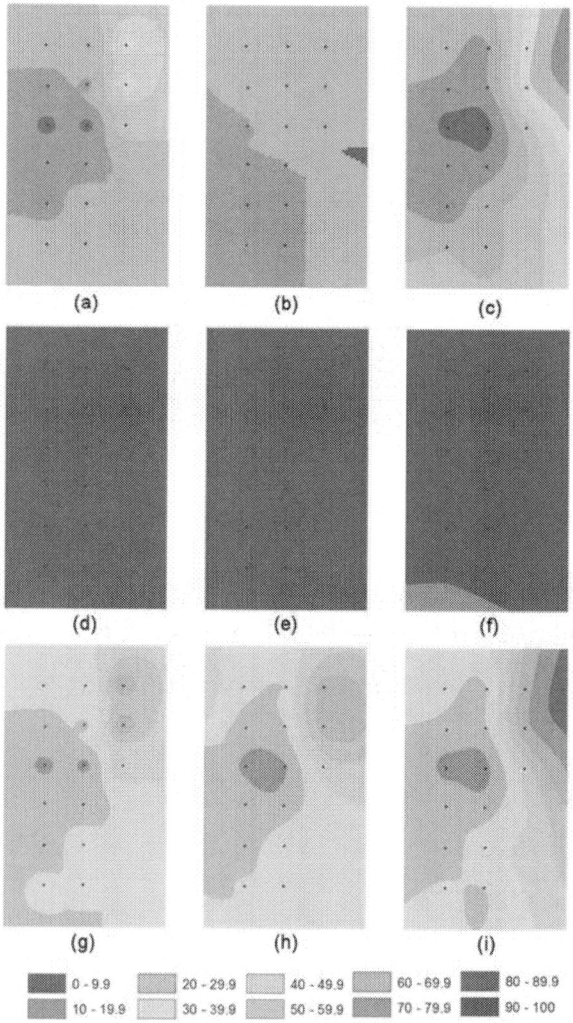

Figure 6. Nine interpolated surfaces for percentage vegetation, impervious surface, and soil, created by IDW, kriging, and spline spatial interpolation methods, respectively. (a) Veg: idw; (b) Veg: kriging; (c) Veg: spline; (d) Imp: idw; (e) Imp: kriging; (f) Imp: spline; (g) Soil: idw; (h) Soil: kriging; (i) Soil: spline.

except two points near the south edge of the study area, which is close to parking lots where impervious surface exists.

Three spatial interpolations were chosen for demonstration purposes. They are: IDW (parameters: power 2, 10 neighbors), spline (parameters: tension, 10 neighbors), and kriging (parameters: circular, 10 neighbors). Each interpolation was applied to create V, I, and S surfaces. In total, there were nine surfaces generated. **Figure 6** shows these nine interpolated surfaces. One may quickly observe how differently these surfaces appear, especially among these vegetation percent surfaces. One may also notice that among three impervious surfaces, only spline surface shows data values greater than 10, which is along the south edge. Among three vegetation surfaces, only spline shows data values in orange or red colors (very low) near the northeast corner. Among three soil surfaces, only spline shows data values in blue colors (very high) near the northeast corner. These are extreme values near edges of interpolated surfaces, a trend associated with spline interpolation, as observed in the earlier examples, also shown in **Figures 2(e)** and **4(b)**, as well as discussed in Ref. [14].

6.2. Evaluation and visualization of spatial interpolation in a closed system

Figure 7 shows these surfaces in 3D visualization, looking from the southeast. **Figure 7(a)** shows the three percentage surfaces generated by IDW, top surface for vegetation, middle surface for soil, and bottom surface for impervious surface. **Figure 7(b)** shows the three percentage surfaces generated by kriging, and **Figure 7(c)** for spline. Bars indicate locations of sampled points. Height of bars equals the percent of vegetation. One may observe that bars poking out of kriging vegetation surface, means an inexact interpolation. One may also observe the extreme data values on spline surfaces. In this 3D visualization, it is evident that three interpolation methods performed very differently.

When adding three surfaces generated by IDW together, because it is a closed system, all cells supposedly should have a data value close to 100%. So do three surfaces generated by kriging and spline. **Figure 8** shows the sum of three surfaces generated by IDW, kriging, and spline. **Figure 8(a)** shows the sum of V, I, and S surfaces generated by IDW. One can

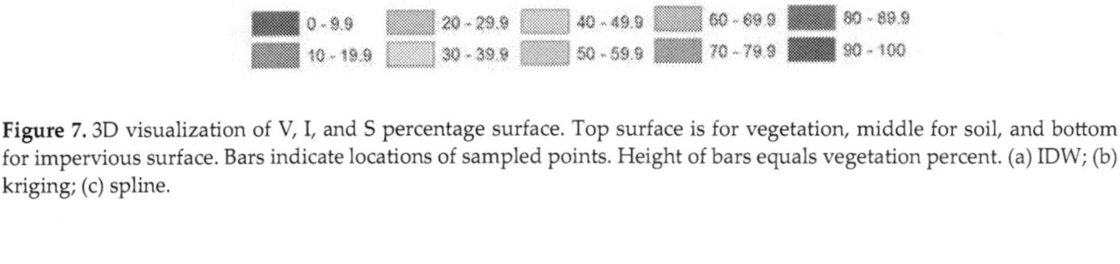

Figure 7. 3D visualization of V, I, and S percentage surface. Top surface is for vegetation, middle for soil, and bottom for impervious surface. Bars indicate locations of sampled points. Height of bars equals vegetation percent. (a) IDW; (b) kriging; (c) spline.

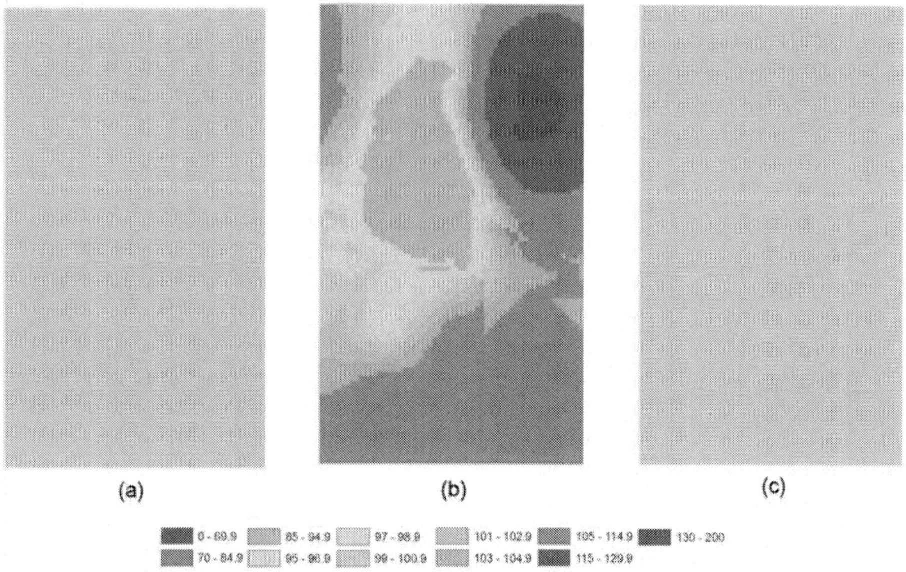

Figure 8. 2D visualization of sum surfaces estimated by (a) IDW, (b) kriging, and (c) spline.

	Min.	Max.	Mean	S.D.
Sum surface estimated by IDW	100	100	100	0
Sum surface estimated by kriging	85.12	132.00	104.68	9.69
Sum surface estimated by spline	100	100	100	0

Table 3. Descriptive statistics for three sum surfaces estimated by IDW, kriging, and spline

observe that there is no major variation from 100% in sum percentage as all cells fall into the category of 99–100.9 range. One may also observe the same trend for spline as shown in **Figure 8(c)**. However, kriging as displayed in **Figure 8(b)** shows a lot of variations from 100% in the sum of V, I, and S percentage. This is another evidence of inexact interpolation, as the interpolated data are not true to the sampled data even at the exact location where it is sampled. **Table 3** shows descriptive statistics for these sum surfaces. One may clearly see that kriging is the only interpolation method that failed to meet the closed system requirement (sum of all variables equals to 100%) when individual variable is interpolated separately and independently.

Figure 9 shows these three sum surfaces in a 3D visualization environment. Bars indicate the locations of sampled data. Height of bars is set at 100, the requirement for a closed system.

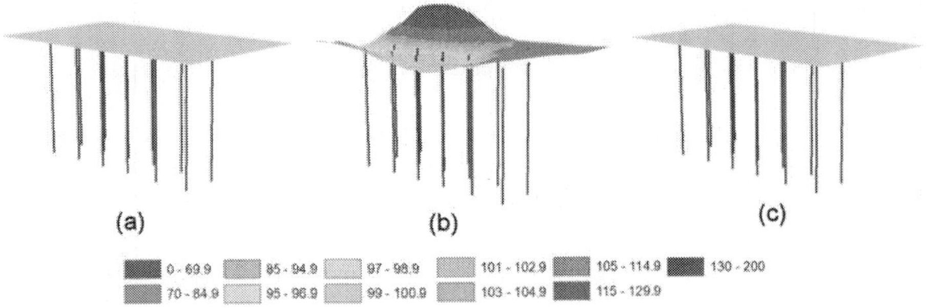

Figure 9. 3D visualization of sum surfaces estimated by (a) IDW, (b) kriging, and (c) spline. Bars indicate location of sampled data, with height set to 100.

One may again observe the bars overreach or underreach the kriging sum surface, an indication of inexact interpolation. On the other hand, IDW and spline seem to quite meet the 100% requirement.

It has to be noted that in this experiment, there are only 15 sample points. It is a very small number of samples. The results in this experiment can be biased due to small sample. Nevertheless, some interesting trends were observed by 2D and/or 3D visualization, which was not evident in quantitative assessment. When examining the interactions between interpolated surfaces in a closed system, both IDW and spline met the requirement, i.e., summing variables to 100%, even though each surface was generated from one variable separately and independently. On the other hand, kriging failed to meet this requirement. It was observed again that kriging is an inexact interpolation. Furthermore, it was also observed that spline, as reported earlier in this study and in Ref. [14], had the tendency to generate extreme values along edges of the study area.

7. Conclusion

In this study, three spatial interpolation algorithms (IDW, kriging, and spline) were applied to a set of soil pH value data to demonstrate the complexity of the process to validate the results of spatial interpolation. Three methods of validation were used: quantitative assessment, 2D visualization, and 3D visualization. Each validation method revealed different characteristics of each spatial interpolation. With quantitative assessment, it was observed that IDW and kriging performed similarly, and both are better than spline. With 2D visualization, it was observed that IDW and spline performed similarly, while kriging performed differently, not necessarily in a good or bad way. With 3D visualization, it was observed that kriging is an inexact interpolation. It was also observed that spline had a tendency to create extreme values along edges of the study area.

Another experiment was conducted to demonstrate the interactions between interpolated surfaces, especially in a closed system. There were three variables in this closed system, each represented a percentage of a specific land cover in an urban area. In a closed system, these three variables should sum up to 100%. Three spatial interpolation algorithms (IDW, kriging, and spline) were applied to each variable separately and independently. These interpolated surfaces were then added up to form a sum surface. It was observed that both IDW and spline successfully met the requirement, making the sum surface 100% for all cells, while kriging failed to meet this requirement.

In conclusion, each spatial interpolation algorithm performed differently. One has to be careful on evaluation of the results. Though quantitative assessment is commonly and widely used in most spatial interpolation studies, it is essential to understand that evaluation of a spatial interpolation should not rely on quantitative assessment alone. 2D visualization and 3D visualization can reveal some facts that cannot be observed in quantitative assessment.

Author details

Yi-Hwa (Eva) Wu* and Ming-Chih Hung

*Address all correspondence to: ywu@nwmissouri.edu

Department of Humanities and Social Sciences, Northwest Missouri State University, University Drive, Maryville, Missouri, USA

References

[1] Longley, P. A., Goodchild, M. F., Maguire, D. J., & Rhind, D. W. Geographic information systems and sciences, 2nd Ed. West Sussex, UK: John Wiley & Sons Ltd.; 2005.

[2] Slocum, T. A., McMaster, R. B., Kessler, F. C., & Howard, H. H. Thematic cartography and geographic visualization, 2nd Ed., Upper Saddle River, NJ, USA: Pearson Prentice Hall; 2005.

[3] Cassman, K. G. Ecological intensification of cereal production systems: Yield potential, soil quality, and precision agriculture. Proceedings of National Academy of Sciences. 1999; 96(11): 5952–5959.

[4] Brase, T. A. Precision agriculture. Clifton Park, NY, USA: Thomson Delmar Learning; 2006.

[5] Adamchuk, V. I., Morgan, M. T., & Loewenberg-Deboer, J. M. A model for agro-economic analysis of soil pH mapping. Precision Agriculture. 2004; 5: 111–129.

[6] Bongiovanni, R., & Lowenberg-DeBoer, J. Economics of variable rate lime in Indiana. Precision Agriculture. 2000; 2: 55–70.

[7] Borgelt, S. C., Searcy, S. W., Stout, B. A., & Mulla, D. J. Spatially variable liming rates: A method for determination. Transactions of the American Society of Agricultural Engineers. 1994; 37: 1499–1507.

[8] Gebbers, R., & Adamchuk, V. I. Precision agriculture and food security. Science. 2010; 327: 828–831.

[9] Heiniger, R. W., & Meijer, A. J. Why variable rate application of lime has increased grower profits and acceptance of precision agriculture in the Southeast. Fifth International Conference on Precision Agriculture Proceedings. Bloomington, MN. ASA, CSSA & SSSA, Madison, WI, USA; 2000.

[10] Environmental Systems Research Institute (ESRI). ArcMap 9.3 ArcGIS desktop help. Redlands, CA: ESRI; 2008.

[11] Bolstad, P. GIS fundamentals: A first text on geographic information systems, 3rd Ed. White Bear Lake, MN, USA: Eider Press; 2008.

[12] Brouder, S. M., Hofmann, B. S., & Morris, D. K. Mapping soil pH: Accuracy of common soil sampling strategies and estimation techniques. Soil Science Society of America Journal. 2005; 69: 427–442.

[13] Corstanje, R., Grunwald, S., Reddy, K. R., Osborne, T. Z., & Newman, S. Assessment of the spatial distribution of soil properties in a northern Everglades marsh. Journal of Environmental Quality. 2006; 35: 938–949.

[14] Wu, Y. H. & -Hung, M. -C. & Patton, J. Assessment and visualization of spatial interpolation of soil pH values in farmland. Precision Agriculture. 2013; 14(6): 565–585.

[15] Watson, M. E., & Brown, J. R. pH and lime requirement. In: J. R. Brown, (Ed.) Recommended chemical soil test procedures for the north central region. Missouri Agricultural Experiment Station, Columbia, Missouri, USA: North Central Regional Publication, no. 221 (Revised); 1998.

[16] Soil Survey Staff. Soil Survey Field and Laboratory Methods Manual. Soil Survey Investigations Report No. 51, Version 2.0. R. Burt and Soil Survey Staff (ed.). U.S. Department of Agriculture, Natural Resources Conservation Service; 2014.

[17] Tobler, W. A computer movie simulating urban growth in the Detroit region. Economic Geography. 1970; 46(2): 234–240.

[18] Schloeder, C. A., Zimmerman, N. E., & Jacobs, M. J. Comparison of methods for interpolating soil properties using limited data. Soil Science Society of America Journal. 2001; 65: 470–479.

[19] Mueller, T. G., Pierce, F. J., Schabenberger, O., & Warncke, D. D. Map quality for site-specific fertility management. Soil Science Society of America Journal. 2001; 65: 1547–1558.

[20] Kravchenko, A. N. Influence of spatial structure on accuracy of interpolation methods. Soil Science Society of America Journal. 2003; 67: 1564–1571.

[21] Lu, G. Y., & Wong, D. W. An adaptive inverse-distance weighting spatial interpolation technique. Computers & Geosciences. 2008; 34(9): 1044–1055.

[22] Bekele, A., Downer, R. G., Wolcott, M. C., Hudnall, W. H., & Moore, S. H. Comparative evaluation of spatial prediction methods in a field experiment for mapping soil potassium. Soil Science. 2003; 168(1): 15–28.

[23] Laslett, G. M., Handcock, M. S., Merier, K., Nychka, D., & Machler, M. B. Kriging and splines: An empirical comparison of their predictive performance in some applications. Journal of the American Statistical Association. 1994; 89(426): 391–409.

[24] Gotway, C. A., Ferguson, R. B., Hergert, G. W., & Peterson, T. A. Comparison of kriging and inverse-distance methods for mapping soil parameters. Soil Science Society of America Journal. 1996; 60: 1237–1247.

[25] Bishop, T. F., & McBratney, A. B. A comparison of prediction methods for the creation of field-extent soil property maps. Geoderma. 2001; 103(1–2): 149–160.

[26] Laslett, G. M., McBrantney, A. B., Pahl, P. J., & Hutchinson, M. F. Comparison of several spatial prediction methods for soil pH. European Journal of Soil Science. 1987; 38(2): 325–341.

[27] Voltz, M., & Webster, R. A comparison of kriging, cubic splines and classification for predicting soil properties from sample information. European Journal of Soil Science. 1990; 41(3): 473–490.

[28] Robinson, T. P., & Metternicht, G. Testing the performance of spatial interpolation techniques for mapping soil properties. Computers and Electronics in Agriculture. 2006; 50(2): 97–108.

[29] Simpson, G, & Wu, Y. -H. Accuracy and effort of interpolation and sampling: Can GIS help lower field costs? ISPRS International Journal of Geo-Information. 2014; 3: 1317–1333.

[30] Kravchenko, A., & Bullock, D. G. A comparative study of interpolation methods for mapping soil properties. Agronomy Journal. 1999; 91: 393–400.

[31] Alagarswamy, G., Boote, K. J., Allen, Jr., L. H., & Jones, J. W. Evaluating the CROPGRO–soybean model ability to simulate photosynthesis response to carbon dioxide levels. Agronomy Journal. 2006; 98: 34–42.

[32] Ridd, M.K. Exploring a V-I-S (vegetation-impervious surface-soil) model for urban ecosystem analysis through remote-sensing—Comparative anatomy for cities. International Journal of Remote Sensing. 1995; 16(12): 2165–2185.

[33] Hung, M. -C. & Ridd, M.K. A sub-pixel classifier for urban land cover mapping based on a maximum likelihood approach and expert system rules. Photogrammetric Engineering and Remote Sensing. 2002; 68(11): 1173–1180.

[34] Germaine, K. & Hung, M. -C. Delineation of impervious surface from multispectral imagery and LiDAR incorporating knowledge-based expert system rules. Photogrammetric Engineering and Remote Sensing. 2011; 77(1): 75–85.

Chapter 3

Wage Concentration in Spain: A Spatial Analysis

Beatriz Larraz, Mónica Navarrete and
José Manuel Pavía

Additional information is available at the end of the chapter

http://dx.doi.org/10.5772/64131

Abstract

In this article, the degree of concentration of wages in Spain at the provincial and regional levels is estimated using the latest available micro-data corresponding to the Structure of Earnings Survey 2010 (N = 216.769). From the analysis of the statistics obtained, it is possible to know in detail the spatial distribution of national wage inequality, to identify those areas where inequality is greatest, and to estimate the possible existence of spatial dependence and structure. The analysis focuses not only on the study of global inequality, but delves into the question by extending the analysis from a gender perspective.

Keywords: inequality, Gini, gender, wages, spatial statistics, variogram

1. Introduction

It is not in dispute that there is growing concern for the increase in inequality in household disposable income in recent decades, both nationally and internationally [1]. In the case of Spain, the report of the International Monetary Fund [2] states that the Gini index [3] on disposable income has risen nearly 3% points, from 31.8 in 1980 to 34.7 in 2010, while in other advanced economies such as the United Kingdom or the United States, the increases were even 6.5 and 7.5% points, respectively, rising from 27.0 to 33.5 in the case of the United Kingdom and 30.1 in 1980 to a more worrying 38.6 in the United States in the same period. Considering the most up-to-date data, in the case of Latin America and the Caribbean zone in 2012, the net Gini average is 44.2 points out of 100, the latest data available for China stand the Gini inequality index at 47.3, reaching in India a value of 47.7 [4].

Wage inequality is found partly in the origins of this income inequality, since, in general, higher wages mean high disposable income and vice versa. From this point of view, the study of inequality in the distribution of wages among the population of a country is especially relevant.

However, besides the importance of dealing with the study of global income inequality in the country, from a sociological perspective and gender, it is equally important to develop indicators of equality between women and men, such as those carried out by some of different women's institutes [5, 6]. It is essential to know which degree of inequality in the distribution of wealth occurs among women themselves and among men themselves, as different groups. Also, in this area of study, the analysis of inequality between women and men, understood as gender inequality, also acquires special relevance. In the case that concerns us as wage concentration, between and within-group index concentration decomposition can be accomplished following the decomposition of Larraz [7].

But as we are dealing with spatially located data, because individuals (employees) work in a specific geographical area, the study of the spatial dependence should be taken into account. To analyze the evidence of existing spatial correlation, the study of this phenomenon should be accomplished from a spatial approach. It seems important to emphasize that the information provided by the spatial location of each observation through its spatial coordinates should not be underestimated. It will be important for spatial modeling, for example.

With this in mind, the aim of this study is to conduct an analysis of the degree of overall, female, male and gender wage concentration at regional and provincial levels (NUTS2 and NUTS3, respectively, [8]) in Spain, as an example that could be replicated in different countries. The study has two purposes: First, to point out the Spanish regions and provinces in which the concentration of wages is higher and more worrying, and, second, to determine the absence or existence of some degree of spatial dependence in the phenomenon. For the first part, this chapter analyzes the maps showing global inequality, inequality among women themselves, among men themselves and gender inequality at regional and provincial levels. To achieve the second objective, the analysis of the spatial dependence of the inequality and its structure has been carried out.

Thus, the chapter is structured as follows. After this introductory section, Section 2 describes the methodology used in the concentration index decomposition and the theory associated with spatial autocorrelation and its structure. Section 3 briefly describes the structure of earnings survey [9] developed by the Spanish National Statistics Institute, whose micro-data [10] have been used to calculate such concentration indices. Section 4 shows the maps, the spatial correlation tests, the estimation of experimental variograms, the fit of theoretical variograms and their analysis to finally conclude with Section 5.

2. Methodology

This section, first, makes reference to the methodology used in the decomposition of the concentration index in between- and within-group components. Subsequently, a brief summary is made of the study of spatial autocorrelation of global, gender, male and female inequality indexes, to conclude with the classic analysis of the structure of spatial dependence in geostatistics.

2.1. Measuring inequality

When addressing the quantitative study of the degree of concentration of an economic variable, the Gini index [3] continues to constitute, after nearly a century of existence, the most used inequality coefficient by official statistical agencies [9, 11] and in the scientific literature [12, 13]. In the case study of income or wages, the said concentration index is based on the relationship between the cumulative proportion of population, $p_i = i/n$, and income, $q_i = A_i/A_n$; where $A_i = \sum_{k=1}^{i} x_k$ and $\{x_i\}_{i=1}^{n}$ represent the individual earnings ordered from the smallest to the largest.

$$IG = \frac{\sum_{i=1}^{n-1}(p_i - q_i)}{\sum_{i=1}^{n-1} p_i}, \forall n \quad (1)$$

The values of this index range from zero, which corresponds to a level of total equal distribution, to one maximum economic concentration or total inequality in the distribution of the variable.

An equivalent expression that returns exactly the same result [7] is based on the definition of mean difference of Gini [14] and is given by the expression

$$IG = \frac{\sum_{i=1}^{n}\sum_{j=1}^{n}|x_i - x_j|}{2\bar{x} \cdot n(n-1)}, \quad \forall n \quad (2)$$

being $\bar{x}$ the arithmetic mean of income.

Despite the validity of these definitions, it should be noted that expression (1) can only be applied if frequencies are unitary, so their use with survey data is limited, as the raising factor (or weight) involves repeating each entry a number of times, not an integer most of the time.

To overcome this limitation, when the available frequencies are not unitary, the calculation of the inequality Gini index should be made through the following expression:

$$IG = \frac{\sum_{i=1}^{n}\sum_{j=1}^{n}\left|x_i - x_j\right| n_i n_j}{2\overline{x}\cdot N(N-1)}, \quad \forall N \tag{3}$$

where N is the total number of individuals and n_i the number of them that has a salary of x_i currency units, which again are arranged in ascending order: $(x_i; n_i)$ and $x_i \leq x_j$ if $i < j$.

When addressing this issue from a gender perspective, in this article, besides the concentration index among all individuals, indexes of male and female concentration are computed, understanding those calculated on all women and men separated. With them it is intended to understand what is happening with the inequalities among women themselves on the one hand and among men on the other hand. These indices will be important to know the reality and decide in which cases are relatively more urgent to take action, if that is the case.

Thus, the rate of female concentration is defined as

$$IG_{\text{Women}} = \frac{\sum_{i=1}^{n_W}\sum_{j=1}^{n_W}\left|x_{Wi} - x_{Wj}\right| n_{Wi} n_{Wj}}{2\overline{x}_W \cdot N_W(N_W - 1)} \tag{4}$$

where N_W is the total number of women and n_{Wi} is the number that has revenues of monetary x_{Wi} units, which again the x_{Wi} are arranged in ascending order: $(x_{Wi}; n_{Wi})$ and $x_{Wi} \leq x_{Wj}$ if $i < j$.

Similarly, the male concentration ratio is defined as

$$IG_{\text{Men}} = \frac{\sum_{i=1}^{n_M}\sum_{j=1}^{n_M}\left|x_{Mi} - x_{Mj}\right| n_{Mi} n_{Mj}}{2\overline{x}_M \cdot N_M(N_M - 1)} \tag{5}$$

where N_M is the total number of men and n_{Mi} the number that has revenues of monetary x_{Mi} units, which again the x_{Mi} are arranged in ascending order: $(x_{Mi}; n_{Mi})$ and $x_{Mi} \leq x_{Mj,}$ if $i < j$.

Also, the gender concentration index is defined, IG_{Gender} (6), as the one that calculates the wage gap exclusively between the wages of men compared to women, not including the differences between women and men, who have already been computed in the above indices (expressions 4 and 5). The index has been calculated after adapting the definitions 3 and 4 in [7] to groups of women and men as

$$IG_{\text{Gender}} = \frac{\Delta_{WM}}{\overline{x}_W + \overline{x}_M} \; being \; \Delta_{WM} = \frac{\sum_{i=1}^{n_W}\sum_{r=1}^{n_M}\left|x_{Wi} - x_{Mr}\right| n_{Wi} n_{Mr}}{N_W N_M} \tag{6}$$

Finally, to identify the contribution of inequality between men and women (gross between: IG_{gb}) and men and women together among themselves (within: IG_w), the degree of total inequality (IG) Larraz [7] decomposition can be used, which is given by:

$$IG = IG_w + IG_{gb} \tag{7}$$

where

$$IG_w = IG_{\text{Women}} \frac{N_W - 1}{N-1} \cdot \frac{B_W}{B_n} + IG_{Men} \frac{N_M - 1}{N-1} \cdot \frac{B_M}{B_n} \tag{8}$$

measures the contribution of inequity between the groups to total index and

$$IG_{gb} = IG_{\text{Gender}} \left(\frac{N_W}{N-1} \cdot \frac{B_M}{B_n} + \frac{N_M}{N-1} \cdot \frac{B_W}{B_n} \right) \tag{9}$$

measures the gross between contribution to total inequality, being $B_n = \sum_{i=1}^{n} x_i n_i$ the total wage bill; $B_W = \sum_{i=1}^{n_W} x_{Wi} n_{Wi}$, the total wage bill perceived by all women; and $B_M = \sum_{i=1}^{n_M} x_{Mi} n_{Mi}$, the total wage bill perceived by all men and making the subscripts *W* and *M* refer to the sample of women and men, respectively.

To end the methodology section devoted to the study of wage inequality, it should be noted that the concentration index used is not affected by changes in scale. This means, for example, the fact that *IG* shows less inequality (higher) in one province against another province would not respond to a relative position of lower (higher) wages in the first province with respect to the second province, if not at a lower relative reality of inequality itself. Also, remember that all defined indexes have their field of variation between 0 and 1, indicating a higher value of the index, increased inequality, and smaller, more equity.

2.2. Analysis of the spatial dependence

To study the possible presence of spatial dependence or correlation on observed variables, two alternative approaches can be implemented, according to data and observations pertaining to territorial units (areas) perfectly defined in space or considered as realizations of a random variable in space.

In the first case, in this article, the Moran's *I*-statistic has been used [15], capable of testing whether the hypothesis of the values obtained from sampling of a random variable are distributed totally random in space, or, on the contrary, there is a significant positive association of similar values between neighboring regions. In its construction, it is necessary to resort to the so-called physical contiguity matrix or spatial weight matrix in which the spatial relationship between each pair of locations is translated, to define the concept of proximity. In this case, it has resorted to the concept of physical contiguity first order used by Moran [16] and

Geary [17], where w_{ij} is unitary if i and j are physically adjacent regions (if provinces have a common border) and zero otherwise. Furthermore, suppose that the products $\left(x_i - \bar{x}\right)\left(x_j - \bar{x}\right)$ are calculated, with $\bar{x}$ as the arithmetic mean of the observations. Then, in the case of positive correlation, these products tend to be positive while in the case of alternation will tend to be negative. The statistics based on this principle was developed by Moran and is defined as follows:

$$I = \frac{n}{S_0} \frac{\sum_i \sum_j (X_i - \bar{X})(X_j - \bar{X})}{\sum_i (X_i - \bar{X})^2} \tag{10}$$

It is written in matrix equals

$$I = \frac{n}{S_0} \frac{(\mathbf{X} - \bar{\mathbf{X}})' \mathbf{W} (\mathbf{X} - \bar{\mathbf{X}})}{(\mathbf{X} - \bar{\mathbf{X}})' (\mathbf{X} - \bar{\mathbf{X}})} \tag{11}$$

being $\mathbf{X} - \bar{\mathbf{X}} = \begin{pmatrix} X_1 - \bar{X} \\ \vdots \\ X_n - \bar{X} \end{pmatrix}$ the column vector of the deviations of the values of the performance in relation to its average. In terms of the statistical moments in the case of normality it is obtained as follows:

$$E_{Norm.}(I) = \frac{-1}{n-1} \text{ and } V_{Norm.}(I) = \frac{n^2 S_1 - n S_2 + 3 S_0^2}{(n^2 - 1) S_0^2} - \frac{1}{(n-1)^2} \tag{12}$$

being $S_0 = \sum_i \sum_j w_{ij}$, $S_1 = \frac{1}{2} \sum_i \sum_j \left(w_{ij} + w_{ji}\right)^2$ and $S_2 = \sum_i \left(w_{i0} + w_{0i}\right)^2$ with $w_{i0} = \sum_j w_{ij}$, $w_{0i} = \sum_j w_{ji}$, $i \neq j$ and n the number of locations, while in the case of randomization are:

$$E_{Aleat.}(I) = \frac{-1}{n-1} \text{ and}$$

$$V_{\text{Aleat.}}(I) = \frac{n\left[\left(n^2 - 3n + 3\right) S_1 - n S_2 + 3 S_0^2\right] - k\left[n(n-1) S_1 - 2n S_2 + 6 S_0^2\right]}{(n-1)(n-2)(n-3) S_0^2} - \frac{1}{(n-1)^2} \tag{13}$$

where $k = m_4 / m_2^2$ is the kurtosis coefficient and being $m_r = \frac{1}{n}\sum_i (x_i - \bar{x})^r$.

Regarding the distribution of the statistic, Cliff and Ord [18] demonstrated that when the sample size is large enough, the Moran *I*-statistic follows a normal standardized asymptotic distribution:

$$\frac{I - E(I)}{\sqrt{V(I)}} \rightarrow N(0,1) \tag{14}$$

Thus, a non-significant value does not reject the null hypothesis of no spatial correlation, while a significant positive value, thereof, inform about the presence of a pattern of positive spatial autocorrelation, i.e., the presence of similar concentration values of the variable X between neighboring regions.

In turn, the representation of Moran's scatterplot is shown in this article. This representation is a visualization technique of spatial effects, providing another useful tool in the analysis of the degree of spatial dependence of a variable. This graph shows in abscissa the values of the x_i variable (normalized or not) and in ordinate the resulting spatial average delay (standardized or not) of variable values in other neighboring locations, i.e., those weighted with a value other than zero in the contiguity matrix. Through this cloud of points is achieved by comparing the value of the variable in a location with variable values in neighboring locations (those considered by the contiguity matrix).

On the other hand, from a statistical perspective, the variogram is the essential tool to analyze the dependence of the observed regionalization ([19, 20], among others). Under the framework of intrinsically stationary random functions, the variogram is defined as

$$\gamma(\mathbf{h}) = \frac{1}{2} V\left[X(\mathbf{s}+\mathbf{h}) - X(\mathbf{s})\right] \tag{15}$$

and shows the evolution of the similarity between the values of the phenomenon under study X and observed locations $\mathbf{s}$ and $\mathbf{s} + \mathbf{h}$, $X(\mathbf{s})$ and $X(\mathbf{s} + \mathbf{h})$ separated by physical distance $\mathbf{h}$. Specifically, a constant variogram for all $\mathbf{h}$ shows no spatial dependence of the phenomenon, while a variogram that presents a non-zero slope near the origin of the coordinates is indicative of the existence of some degree of spatial dependence. In this study, it was decided to allocate the concentration index of each province to the spatial coordinates of the capital, where the most of the population is concentrated in general terms.

Although the classical estimator $\gamma(\mathbf{h})$, which is given by the expression (16) [21], is commonly used

$$\gamma^{*}(\mathbf{h}) = \frac{1}{2N(\mathbf{h})} \sum_{i=1}^{N(\mathbf{h})} \left(X(\mathbf{s}_i + \mathbf{h}) - X(\mathbf{s}_i) \right)^2 \tag{16}$$

(being, in our case, $X(\mathbf{s}_i)$ the value corresponding to inequality index in the province whose provincial capital has coordinates $\mathbf{s}_i$ and $N(\mathbf{h})$ the number of pairs of provinces whose capitals are separated by a vector $\mathbf{h}$), this chapter has chosen to use the estimator by Cressie and Hawkins [22], given by equation (17), due to its larger robustness (see [23] for different options).

$$\gamma^{*}_{CH}(\mathbf{h}) = \frac{1}{2}\left[0.457 + \frac{0.494}{N(\mathbf{h})}\right]^{-1}\left[\frac{1}{N(\mathbf{h})}\sum_{i=1}^{N(\mathbf{h})} \left| X(\mathbf{s}_i + \mathbf{h}) - X(\mathbf{s}_i) \right|^{\frac{1}{2}}\right]^{4} \tag{17}$$

3. Data from the structure of earnings survey

To carry out the study of wage inequality, we used the latest available information: micro-data on the distribution of wages, included in the Structure of Earnings Survey 2010 [9], with a 4-year basis. Through its micro-data, you can have, besides sex, individual wages and the province (NUTS3) in which they perform their job, data that will allow us to carry out the proposed study. This information is gathered from the quotation centers selected in the sample,

Figure 1. Map of Spain at the provincial level [8]. Source: own elaboration.

http://dx.doi.org/10.5772/64131

excluding enterprises in agriculture and fishing, public administration employees not covered by social security, domestic staff and extraterritorial body workers.

The study has taken into account, on the one hand, the variable "gross annual earnings per worker", including payments in goods, to study the different degrees of concentration existing in annual gross wages. The analysis of inequality in the distribution of this variable allows us to analyze the possible consequences to the fact that women have lower gross annual salary on average than men. On the other hand, to isolate the effect on annual salaries increased by the presence of women in part-time jobs (a lower annual average wage), the study was also conducted on the variable "hourly earnings per worker".

The latest survey, dated October 2010, gathers information for about 25,104 quotation centers and 216,769 employees of this, the grossing up factor (weight) is the number of workers in the population corresponding to that information. These workers develop their work in a particular province (NUTS3), which are in turn included in its region (NUTS2) (see **Figures 1** and **2**).

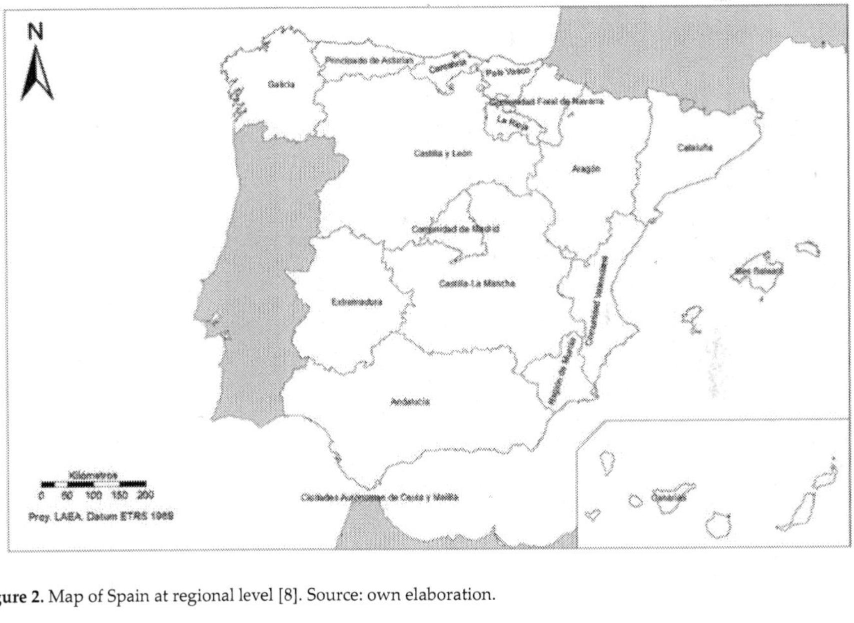

Figure 2. Map of Spain at regional level [8]. Source: own elaboration.

About the most important global statistics of the survey, following **Table 1**, it can be highlighted that though the total population of workers in Spain in 2010 is almost 12 million people, there are still more men (53.2%) than women (46.8%) working. But more important, women as a whole earn just the 40.5% of total annual payroll, while men as a whole earn the rest. Moreover, there is a great difference between the average wages per year of both genders: while men earn

25,479€ per year on average, women earn just 19,735€/year. Measuring this difference on wages per hour, it can be concluded that the gender pay gap is almost 14%.

	Individuals	Total annual payroll (%)	Average wage per year	Total payroll per hour (€/h)	Average wage per hour
Women	5,618,100 (46.8%)	40.5%	19,735.22€	42.7%	10.15€
Men	6,381,446 (53.2%)	59.5%	25,479.74€	57.3%	11.78€
Total	11,999,546		22,790.20€		11.06€
Gender pay gap[1]			22.55%		13.84%

Source: own elaboration from the Structure of Earnings Survey Micro-Data [9].
[1] 'Gender pay gap' indicator is defined by Eurostat as the difference between the average gross hourly earnings of men and women expressed as a percentage of the average gross hourly earnings of men.

Table 1. Descriptive statistics of wage distribution by gender in Spain.

Inequality indicators	
Gini index (on anual salaries)	32.72%
Female Gini index	33.53%
Male Gini index	30.91%
Gender Gini index	33.57%
Low pay rate (1)	13.42%
Proportion of women among total workers with low pay jobs	66.0%
D9/D5 (ninth decile (2) divided by the median (3) of wage per hour)	2.12
D5/D1 (the median divided by the first decile of wage per hour)	1.58
D9/D1 (ninth decile (2) divided by the first decile of wage per hour)	3.34

Source: Spanish Statistics Institute (2012) and own elaboration from the Structure of Earnings Survey micro-data [9].
(1) Proportion of workers whose wage per hour is below two-third of the median salary.
(2) Deciles are the values of the pay that, ordered from smallest to largest, divide the number of workers into 10 equal parts, such that within each are included 10% of them.
(3) Median is the value of the pay that divides the number of workers into two equal parts: the one who have a higher salary and the one who have a lower salary.

Table 2. Inequality indicators of wage distribution in Spain.

Having said that, as if these facts were not serious enough, **Table 2** reports the most important inequality indicators. It is worth noting that female inequality is higher than male one, which means that inequality among women themselves is higher than among men themselves. But it is also of concern that gender inequality, measured by expression (6), is even higher. Moreover, among workers with lower salaries, women account for 66% of them. It is also important to say that the 10% of workers with the highest salaries earn more than double that of median salary and more than three times the salary of the 10% of workers with lowest salary per hour. **Table 3** reports such percentiles.

	Total population	Female	Male
10th percentile	8,643.66	7,001.27	11,391.44
25th percentile	13,602.53	11,462.43	15,770.16
50th percentile	19,017.09	16,536.10	21,206.99
75th percentile	28,255.45	24,751.72	31,164.24
90th percentile	40,811.42	36,249.13	44,725.21

Source: Spanish Statistics Institute (2012).

Table 3. Percentiles by gender (€/year) in Spain.

4. Results

From the micro-data of the Structure of Earnings Survey [9], concentration indices have been calculated: overall, gender, male and female as it has been detailed in Section 2 at regional and provincial level in Spain. This is intended, first, to compare the situation of each region and province regarding the other as far as inequality of wages (annual and hourly) to each of the four concepts. To do this, remember that a higher index value corresponds to a more unequal distribution of variable, whereas a lower value corresponds to a more equitable distribution of the same. This information was reflected in the following maps (**Figures 3** to **6**), showing the detailed information in **Tables A1** and **A2** of the Annex.

The grey scale map has been made by compiling from the distribution decile values of the four indices (not of each separately, so they match the legend of the 4 maps on each figure) to be able to perform a comparison at a glance of the different degrees of inequality obtained. First it is commented at a regional level (NUTS2) to move later at the provincial level (NUTS3).

As aforementioned, it has also carried out the spatial analysis of the different degrees of concentration studied, to know whether a particular pattern of spatial autocorrelation should be considered in future studies.

Thus, in regard to the space study performed from gross annual salary (**Figure 3**)—remembering that it implies the fact that the average annual salary for women is less than the average for men due in part to increased female employment in part-time jobs—at first sight, in general, greater wage concentration of gender in Spain (**Figure 3b**) than global (**Figure 3a**) is observed, i.e., the distribution of wages among workers of different sex is more unequal if one considers all workers together. In addition, inequality among women (**Figure 3c**) is clearly higher than among men (**Figure 3d**) in all the regions (NUTS2).

Specifically, Madrid and Andalusia are positioned as the regions with highest overall wage inequality in Spain, followed by Murcia, Ceuta and Castilla Leon. At the other extreme, the Balearic Islands, Castilla-La Mancha and Navarra present a lower degree of wage concentration. In regard to gender inequality, Murcia, Andalusia, Madrid, Castilla Leon, Ceuta and Cataluña are the regions with highest degree of concentration, continuing Balearic Islands,

Navarra and Castilla-La Mancha as the most equalitarian regions in the distribution of wages between genders.

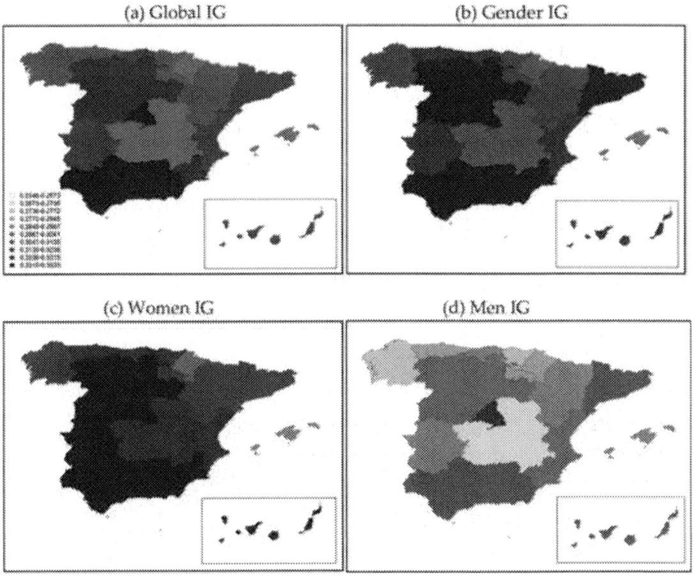

Figure 3. Map of Spain for the concentration index of the entire population (a), gender (b), the group of women (c) and the group of men (d) at regional scale (NUTS2), calculated from the annual gross salary received by each worker. Source: own elaboration.

If the study is carried out within the group of women, it is also of concern that the degree of internal concentration exceeds the generally assumed because of inequality between the distribution of annual wages between women themselves. Specifically, it is in the Midwest and southern Spain (Murcia, Extremadura, Andalusia, Melilla, Castilla Leon, Ceuta, the Canary Islands and Madrid) where this inequality is greater, while Baleares and Navarra regions stand out as having less difference in women's wages.

In the case of male inequality, Madrid is the region where the distribution is less equitable between them, followed by Ceuta, Cataluña and Andalusia, with Castilla-La Mancha, Galicia and the Basque Country more equalitarian.

By correcting the effect of the increased presence of women in part-time jobs on the annual profit by considering the variable of gain/time, we see, first, how wage differences are smoothed in all groups (**Figure 4**), while still maintaining the worrying situation of greater gender than overall concentration, lighter than in the previous case. In general, except in Castilla Leon and Extremadura, less concentration of female than male is now seen. Therefore, regardless of the number of hours worked per worker, data reflect a lower equity in the distribution of hourly wages among men than among women.

Specifically, Madrid and Cataluña now appear as the most unequal regions (**Figure 4a**) together with Castilla Leon, La Rioja and Navarra being the most equitable. In the study of gender (**Figure 4b**), Madrid is the region with the highest degree of concentration between men and women, followed again by Castilla Leon and Cataluña, which are now joined by Asturias and Murcia. On the opposite side, again positioned are La Rioja and Navarra regions where also the distribution of wages between men and women is more equal.

In the case of female wage concentration (**Figure 4c**), making it through the hourly wage implies greater equity in regions such as Asturias, Basque Country, Navarra, La Rioja, Aragon, Valencia, Balearic Islands and Andalusia. At the other end, still stands Castilla Leon with an uneven distribution of the wage mass, in this case between women themselves.

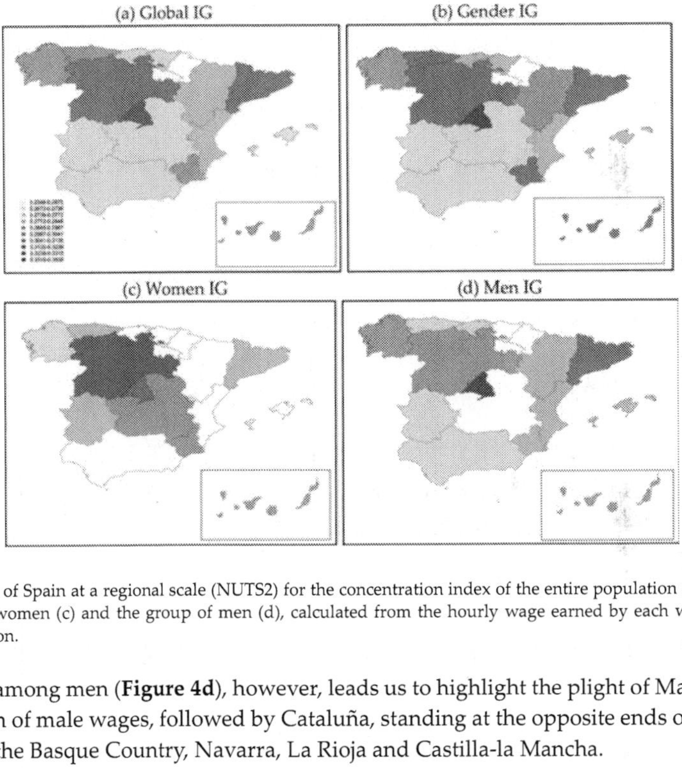

Figure 4. Map of Spain at a regional scale (NUTS2) for the concentration index of the entire population (a), gender (b), the group of women (c) and the group of men (d), calculated from the hourly wage earned by each worker. Source: own elaboration.

The study among men (**Figure 4d**), however, leads us to highlight the plight of Madrid on the distribution of male wages, followed by Cataluña, standing at the opposite ends of equity the regions of the Basque Country, Navarra, La Rioja and Castilla-la Mancha.

In **Figure 3**, a breakdown at the provincial level (NUTS 3), it is noted, first, that the range of values of the concentration indices are higher than the automatic scale. While the regional study on the minimum was 0.2346 and the maximum was 0.3535, in this case we find values between 0.2018 and 0.3868. This implies that the values that correspond to each color range of grays have changed substantially.

That been said, in the analysis of the overall annual wage concentration in the whole country at a provincial level (**Figure 5a**), three problematic areas are observed for their high levels of inequality compared to other provinces, which are western Andalusia and the area in the center and east of the peninsula. In the first stand Huelva, Seville and Cadiz as the provinces with the highest concentration indices. They are followed by Valladolid and Palencia as provinces with high levels of inequality in the distribution of annual wages, next is Madrid. Finally, the provinces of Murcia and Lleida have the highest income inequality in the east of the country, although any of the other three provinces of Cataluña also have high levels of inequality. On the other hand, among the provinces with greater equity in the distribution include Ciudad Real, Castellon, Albacete, Huesca and Alava.

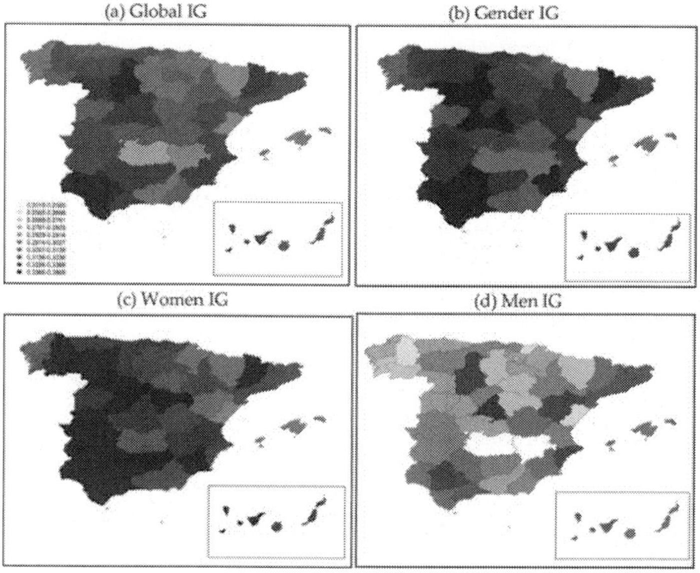

Figure 5. Map of Spain at a provincial level (NUTS3) for the concentration index of the entire population (a), gender (b), the group of women (c) and the group of men (d), calculated from the annual salary received by each worker. Source: own elaboration.

From the map of gender inequality (**Figure 5b**) we can conclude that, overall, these values are higher than the global inequality in each province, being the provinces with the highest levels of concentration gender the same as globally plus Zamora, Avila, Cordoba and Murcia. The areas of greater gender equity are found in Ciudad Real, Huesca, Castellón, Albacete and Baleares Islands.

In general, women's inequality is seen clearly higher than the male, having increased wage inequality among women in the western half of the country (including the two provinces of the Canary Islands) in the East with the exception of Murcia and Lleida, which also have high levels of concentration in the female rate (**Figure 5c**). Standing as provinces with the lowest female inequality are Castellon, Balearic, Navarra, Huesca, Teruel and Soria.

As shown in **Figure 5d**, the degree of concentration in the wages of men seems much more encouraging in general, existing provinces such as Ciudad Real and Albacete with high levels of equity; however Madrid is still of concern, which has the highest degree of male inequality in the country, followed by Sevilla, Alicante, Palencia, Teruel and Valladolid.

Figure 6 shows, as happened at regional level, obviously that the hourly wage distribution is more equal at the provincial level that the distribution of annual salary, since the use of the variable earning per hour involves no consideration of part-time sessions, both male than female.

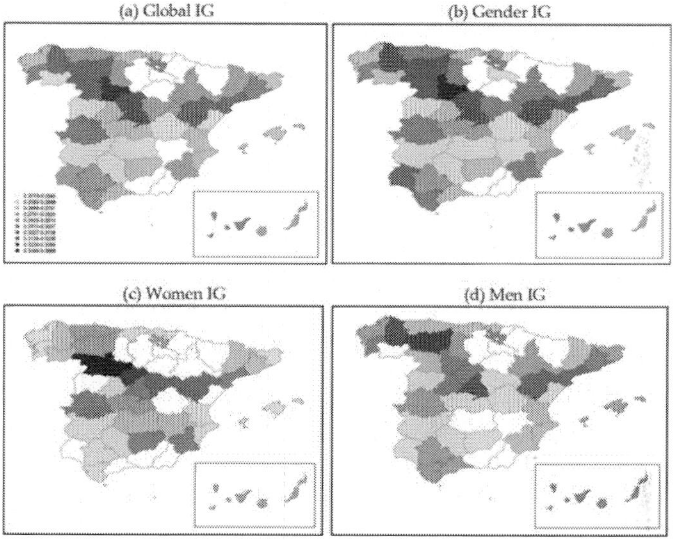

Figure 6. Map of Spain at a provincial level (NUTS3) for the concentration index of the entire population (a), gender (b), the group of women (c) and the group of men (d), calculated from the hourly wage earned by each worker. Source: own elaboration.

In regard to global inequality measured in terms of hourly wages, **Figure 6a** shows two areas in the peninsula clearly with more equitable distributions, such as in the north, the provinces of Soria, Burgos, Navarra, Huesca, La Rioja and Gipuzkoa and, in the southeast, Almería, Granada and Albacete. On the other hand, the greatest inequality is observed in Valladolid, followed by Ceuta, and around by Segovia, Madrid, Leon and Lugo.

Also in the case of the study of hourly earnings, all provinces have greater gender inequality (**Figure 6b**) than overall, although we are talking of figures about a tenth on average. Therefore, although the map shows an image similar to that of global inequality, it is worthy to highlight it for its high concentration the province of Valladolid, followed behind by the provinces of Lugo, Zamora, Ceuta, Segovia, Tarragona, Madrid and Teruel.

	Global IG	Women IG	Men IG	Gender IG
p-Valor annual salary	0.00023***	0.00019***	0.00122***	0.00270***
p-Valor per hour salary	0.01319**	0.16276	0.00789***	0.01069**

Note: *** They refer to the statistical significance of the 1% and ** to 5%.
Source: own elaboration.

Table 4. Test of significance of Moran's *I*-statistic: *p*-values of Moran's *I*-statistic for overall concentration indices, female, male and gender indices.

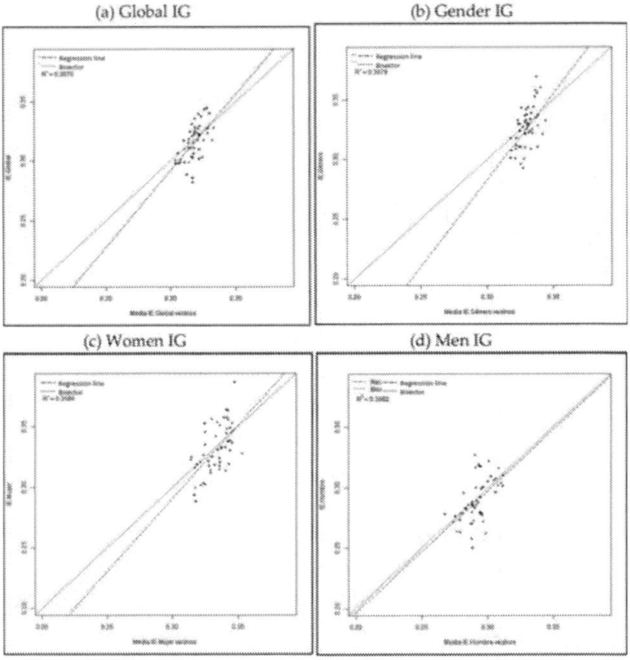

Figure 7. Moran's Scatterplot corresponding to the degree of concentration of the entire population (a), gender (b), the group of women (c) and the group of men (d), calculated from the annual salary perceived by each worker. Source: own elaboration.

Finally, female inequality, shown in **Figure 6c**, shows a clear strip of greater inequality in the center, highlighting Valladolid and Zamora as those provinces with the highest rate of female concentration, followed by Segovia, Ceuta, Melilla and Teruel. Especially equitable for women is the distribution of hourly wages in the North (interior), some provinces in central and

http://dx.doi.org/10.5772/64131

southern Spain. Male inequality (**Figure 6d**) has not yet clear spatial/geographic patterns, highlighting Leon, Lugo, Madrid, Ceuta and Tarragona as provinces with less equity and, at the other end, Soria, Guipuzkoa, and Burgos and Huesca as provinces with more equity in the wage distribution.

Once analyzed the maps of global inequality, gender, male and female in Spain, next a spatial autocorrelation analysis of the phenomenon is performed. First the contrast of lack of spatial correlation to positive existence of spatial correlation is performed, through statistical Moran's *I* (**Table 4**). This statistic is significantly positive in all cases when the variable annual gross salary calculates the indices. In the case of earnings per hour, only in the event of the women's concentration index, the statistic is not statistically significant, in which case one could not reject the hypothesis of no spatial correlation.

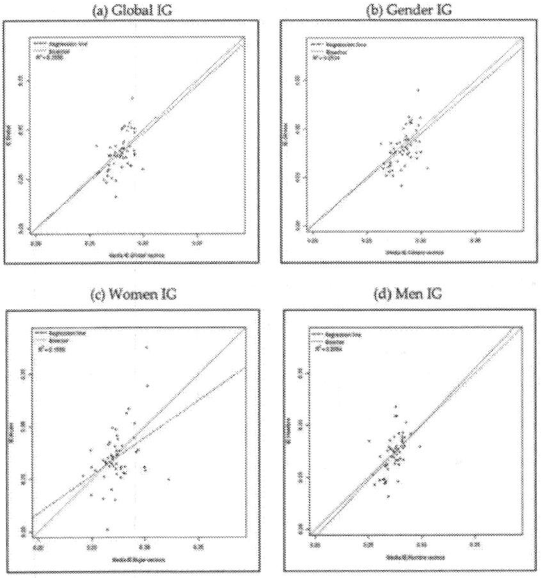

Figure 8. Moran's Scatterplot corresponding to the degree of concentration of the entire population (a), gender (b), the group of women (c) and the group of men (d), calculated from salary when perceived by each worker. Source: own elaboration.

These results can be checked visually in **Figures 7** and **8**, showing the Moran's scatterplot of the concentration indices calculated from the variable annual gross profit and earnings per hour, respectively. **Figure 7** shows how neighboring provinces with high concentration values also and vice versa surrounds provinces with high concentration values. However, in the female concentration index computed from the wage by time, as shown in **Figure 8c**, the point cloud shows the absence of spatial correlation, indicated by the low ratio between the concentrations of the provinces and those of their neighbors.

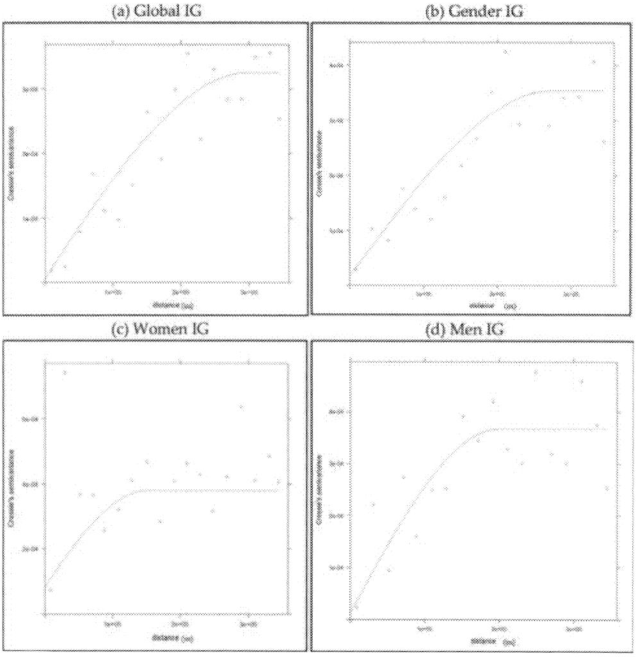

Figure 9. Structural analysis of spatial dependence: Experimental and theoretical variograms adjusted for the concentration index of the entire population (a), gender (b), the group of women (c) and the group of men (d), calculated from gross annual salary received by each worker. Source: own elaboration.

	Global IG		Gender IG	
Model	Sill	Range	Sill	Range
Nugget effect	0.0000061	0	0.0000175	0
Spherical	0.0003189	300.011 km	0.0003364	275.015 km
	Women IG		**Men IG**	
Model	Sill	Range	Sill	Range
Nugget effect	0.0000833	0	0.0000120	0
Spherical	0.0002962	149.976 km	0.0003547	200.003 km

Source: own elaboration.

Table 5. Nested variogram theoretical models, with their sills and ranges used for fitting theoretical variograms of global inequality, gender, male and female calculated from the variable annual gross salary.

Theoretical models	Parameters for global IG	
Model	Sill	Range
Nugget effect	0.0000038	0
Spherical	0.0002718	299.995 km

Source: own elaboration.

Table 6. Theoretical models of the nested variograms, with their sills and ranges used for the adjustment of the directional variograms of global inequality

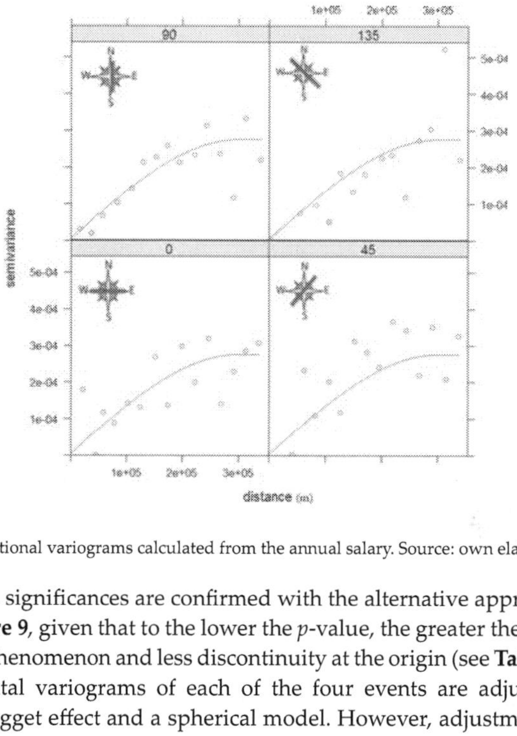

Figure 10. Global IG directional variograms calculated from the annual salary. Source: own elaboration.

The aforementioned significances are confirmed with the alternative approach of variograms represented in **Figure 9**, given that to the lower the *p*-value, the greater the structure of spatial dependence of the phenomenon and less discontinuity at the origin (see **Table 5**). It can be seen how the experimental variograms of each of the four events are adjusted from a linear combination of a nugget effect and a spherical model. However, adjustment parameters (sill and range) are different.

Large discrepancies are observed between them, and while the overall concentration index has a range around 300 km (which is, more or less, the distance from the center of the Iberian Peninsula to the coast), the gender goes down to 275 km, with the male and female of 200 km and 150 km, respectively. Recalling that the lower the range, the smaller the distance where the spatial correlation fades away, in the case of female concentration index, the positive spatial correlation disappears already at short distances. However, the other three cases (the global, male and gender concentration indices) show important spatial correlation structures.

	Global IG		Gender IG	
Model	Sill	Range	Sill	Range
Nugget effect	0.0002142	0	0.0001816	0
Spherical	0.0001240	99.982 km	0.0001886	99.978 km
	Women IG		**Men IG**	
Model	Sill	Range	Sill	Range
Nugget effect	0.0004623	0	0.0001941	0
Spherical	–	–	0.0001777	100.004 km

Source: own elaboration.

Table 7. Theoretical models of the nested variograms, with their sills and ranges used for fitting theoretical variograms of global inequality, gender, male and female calculated from the variable hourly wage.

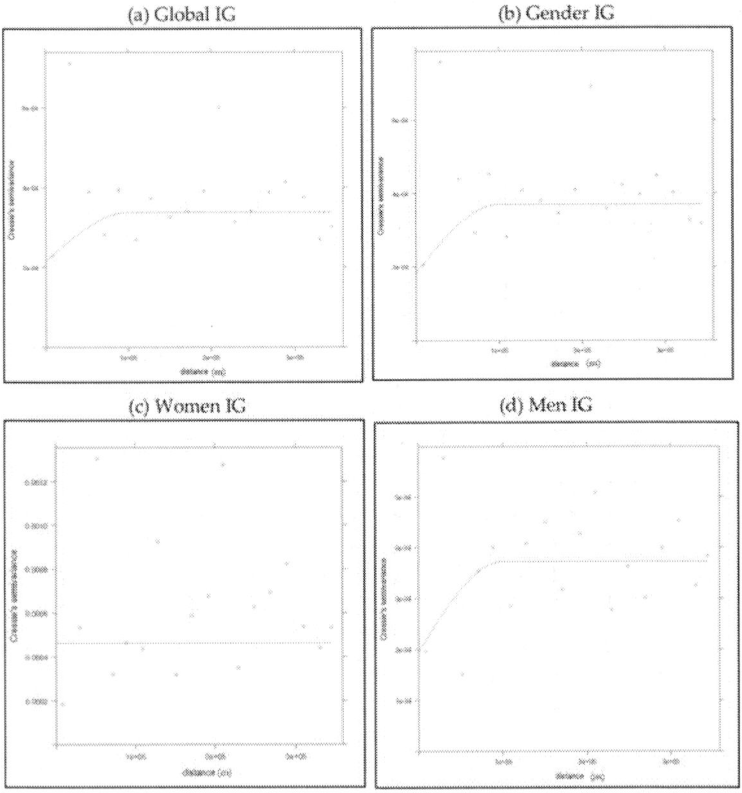

Figure 11. Experimental and theoretical variograms fitted for the concentration index of the entire population (a), gender (b), the group of women (c) and the group of men (d), calculated from hourly wage earned by each worker. Source: own elaboration.

http://dx.doi.org/10.5772/64131

In the case of the global concentration index, directional variograms have also been calculated, to see which one of the main directions presents the greater correlation (**Table 6**). **Figure 10** shows how the fit of the north–south direction experimental variogram is substantially better than in the other three main directions of space.

If we perform the same variogram analysis of the concentration indices calculated from the variable "earnings per hour", in cases of global, gender and male inequality a lower spatial correlation structure is observed than in the case of annual wage, as the ranges are very small (100 km) and the nugget effects are very large in relation to the total variability of the respective processes (**Table 7** and **Figure 11**). In the case of the concentration index among women, the high variability of the process is reflected in the adjustment through a model of pure nugget effect, indicating the absence of spatial correlation (**Figure 11c**).

Consequently, a clear structure of spatial correlation is observed, following a spherical model, in all inequality indices calculated from "annual gross earnings" but very little in the case of global, male and gender indices based on the "gain per hour" and even non-existent in the case of female concentration.

5. Conclusions

In this chapter, the spatial structure of wage inequality in Spain has been analyzed. The study contributes to the literature on wage concentration and gender equality analysis of the various degrees of concentration of regional and provincial areas in Spain. The research was carried out from the last micro-data of the Structure of Earnings Survey conducted by the Spanish Statistical Institute in 2010, both for the whole population and the group of women together, men together and including both genders. This is intended to give pause for the thought on the importance of the information provided by the spatial coordinates of the data and of the growing wage concentration in the hands of a part of society, from a gender perspective.

From this study we conclude, first, that the rate of wage concentration in the group of women are always higher than in the group of men, implying greater inequality in the distribution of wages among female workers than among the men of each region or Spanish province. In addition, in general, a greater gender wage inequality is observed, than global one, worrying fact, which hinders the equality between women and men.

In addition, as expected, a higher degree of concentration in the study of the annual gross profit is observed, than in gross profit per hour, because part-time jobs result in a lower annual salary, thereby increasing the concentration of wage levels.

From a regional perspective, at the top of the table, one corresponding to higher concentration values are Murcia, Madrid, Andalusia, Castilla Leon, Ceuta and Cataluña, joined by Extremadura and Canary Islands in the case of high levels of female inequality. At the bottom of the table, regions such as Balearic Islands, Navarra, Castilla-La Mancha, Aragón and Galicia are highlighted as regions with the lowest rates of wage concentration.

Moreover, this chapter focuses on the study of the spatial analysis of wage concentration indices. We appreciate the presence of positive spatial autocorrelation in the case of the indices calculated by the variable annual gross earnings and earnings per hour, being unable to reject space randomization in the case of female inequality from the earnings per hour.

Finally, through the analysis of the structure of spatial dependence of the phenomenon, greater spatial correlation structure is concluded in the indices calculated on the annual profits, which have ranges of up to 300 km and small nuggets effects, while those calculated on earnings per hour see its ranges reduced to 100 km and present indicative nugget effect of lower spatial correlation at smaller distances. Even, the female concentration measured in hourly earnings must have been modeled through a pure nugget effect, showing the absence of spatial correlation. Such patterns of spatial autocorrelation present in wage inequality should be considered in future studies.

Annex

	Annual wage				Per hour wage			
Province	**IG**	**IG_{Woman}**	**IG_{Man}**	**IG_{Gender}**	**IG**	**IG_{Woman}**	**IG_{Man}**	**IG_{Gender}**
Araba/Álava	0.2977	0.3162	0.2766	0.3074	0.2833	0.2768	0.2846	0.2845
Albacete	0.2881	0.3252	0.2576	0.3005	0.2575	0.2695	0.2478	0.2608
Alicante/Alacant	0.3243	0.3177	0.3194	0.3300	0.2735	0.2647	0.2751	0.2761
Almería	0.3033	0.3145	0.2847	0.3102	0.2457	0.2306	0.2459	0.2516
Ávila	0.3273	0.3637	0.2788	0.3435	0.2752	0.2609	0.2866	0.2752
Badajoz	0.3175	0.3485	0.2860	0.3250	0.2652	0.2686	0.2612	0.2662
Balears (Illes)	0.2979	0.2937	0.2956	0.3010	0.2735	0.2673	0.2769	0.2742
Barcelona	0.3213	0.3213	0.3062	0.3305	0.2874	0.2728	0.2894	0.2922
Burgos	0.3025	0.3141	0.2731	0.3222	0.2480	0.2298	0.2437	0.2572
Cáceres	0.3152	0.3509	0.2857	0.3242	0.2830	0.2874	0.2787	0.2844
Cádiz	0.3400	0.3572	0.3015	0.3602	0.2776	0.2592	0.2800	0.2833
Castellón/Castelló	0.2856	0.2884	0.2618	0.2998	0.2654	0.2514	0.2668	0.2704
Ciudad Real	0.2821	0.3135	0.2501	0.2923	0.2591	0.2710	0.2473	0.2619
Córdoba	0.3293	0.3633	0.2992	0.3387	0.2638	0.2644	0.2619	0.2648
Coruña (A)	0.3007	0.3119	0.2780	0.3087	0.2736	0.2618	0.2791	0.2760
Cuenca	0.3060	0.3183	0.2923	0.3096	0.2616	0.2555	0.2662	0.2616
Girona	0.3176	0.3203	0.3094	0.3219	0.2718	0.2593	0.2770	0.2733
Granada	0.3017	0.3191	0.2786	0.3085	0.2506	0.2397	0.2578	0.2511
Guadalajara	0.3110	0.3524	0.2723	0.3282	0.2803	0.2971	0.2619	0.2886
Gipuzkoa	0.3005	0.3248	0.2634	0.3173	0.2569	0.2610	0.2427	0.2646
Huelva	0.3445	0.3481	0.2962	0.3692	0.2809	0.2541	0.2679	0.3002
Huesca	0.2885	0.3017	0.2707	0.2955	0.2531	0.2582	0.2443	0.2569

http://dx.doi.org/10.5772/64131

Province	Annual wage				Per hour wage			
	IG	IG_{Woman}	IG_{Man}	IG_{Gender}	IG	IG_{Woman}	IG_{Man}	IG_{Gender}
Jaén	0.3236	0.3571	0.2857	0.3366	0.2725	0.2847	0.2601	0.2751
León	0.3225	0.3447	0.2881	0.3352	0.3023	0.2773	0.3178	0.3039
Lleida	0.3284	0.3452	0.3004	0.3377	0.2761	0.2724	0.2740	0.2789
Rioja (La)	0.3066	0.3228	0.2792	0.3177	0.2556	0.2527	0.2523	0.2588
Lugo	0.3171	0.3418	0.2636	0.3329	0.3017	0.2729	0.3087	0.3122
Madrid	0.3357	0.3333	0.3270	0.3419	0.3031	0.2888	0.3085	0.3059
Málaga	0.3275	0.3383	0.3078	0.3370	0.2714	0.2549	0.2777	0.2737
Murcia	0.3311	0.3535	0.3000	0.3431	0.2814	0.2825	0.2750	0.2849
Navarra	0.2982	0.2997	0.2785	0.3117	0.2490	0.2346	0.2468	0.2557
Ourense	0.3112	0.3526	0.2737	0.3216	0.2628	0.2757	0.2455	0.2696
Asturias	0.3109	0.3238	0.2823	0.3255	0.2804	0.2766	0.2688	0.2894
Palencia	0.3378	0.3321	0.3177	0.3557	0.2787	0.2576	0.2773	0.2866
Palmas (Las)	0.3173	0.3305	0.3012	0.3209	0.2807	0.2766	0.2832	0.2811
Pontevedra	0.2989	0.3173	0.2716	0.3103	0.2829	0.2615	0.2880	0.2875
Salamanca	0.3031	0.3277	0.2797	0.3101	0.2596	0.2490	0.2663	0.2592
Santa Cruz de Tenerife	0.3278	0.3367	0.3104	0.3329	0.2856	0.2806	0.2847	0.2884
Cantabria	0.3153	0.3306	0.2863	0.3293	0.2708	0.2614	0.2695	0.2754
Segovia	0.3257	0.3561	0.2875	0.3336	0.3037	0.3172	0.2851	0.3074
Sevilla	0.3437	0.3556	0.3218	0.3575	0.2778	0.2642	0.2796	0.2810
Soria	0.2986	0.3033	0.2707	0.3209	0.2314	0.2018	0.2310	0.2408
Tarragona	0.3221	0.3186	0.3053	0.3331	0.2993	0.2912	0.2929	0.3065
Teruel	0.3210	0.3024	0.3177	0.3311	0.3001	0.2981	0.2919	0.3053
Toledo	0.3098	0.3331	0.2772	0.3234	0.2736	0.2810	0.2627	0.2775
Valencia/València	0.3182	0.3350	0.2898	0.3294	0.2725	0.2620	0.2740	0.2760
Valladolid	0.3396	0.3467	0.3165	0.3527	0.3316	0.3746	0.2883	0.3399
Bizkaia	0.3101	0.3264	0.2832	0.3234	0.2736	0.2662	0.2697	0.2790
Zamora	0.3236	0.3868	0.2798	0.3442	0.3012	0.3386	0.2714	0.3117
Zaragoza	0.3113	0.3211	0.2832	0.3274	0.2769	0.2523	0.2769	0.2856
Ceuta	0.3279	0.3377	0.3161	0.3371	0.3069	0.3124	0.3025	0.3094
Melilla	0.3235	0.3446	0.3039	0.3285	0.2970	0.3027	0.2915	0.2982
Spain	**0.3272**	**0.3353**	**0.3091**	**0.3357**	**0.2879**	**0.2779**	**0.2895**	**0.2910**

Source: own elaboration.

Table A1. Provincial concentration indices on the whole population (IG), on the group of women (IG_{Women}), men (IG_{Men}) and gender (IG_{Gender}) concentration index. All have been calculated from the variable annual gross earnings and earnings per hour.

Autonomic community	Annual wage				Per hour wage			
	IE	IE_{Woman}	IE_{Man}	IE_{Gender}	IE	IE_{Woman}	IE_{Man}	IE_{Gender}
Andalucía	0.3315	0.3483	0.3060	0.3430	0.2708	0.2732	0.2596	0.2735
Aragón	0.3092	0.3160	0.2861	0.3224	0.2762	0.2753	0.2588	0.2834
Asturias	0.3109	0.3238	0.2823	0.3255	0.2804	0.2688	0.2766	0.2894
Baleares	0.2979	0.2937	0.2956	0.3010	0.2735	0.2769	0.2673	0.2742
Canarias	0.3229	0.3348	0.3058	0.3271	0.2839	0.2842	0.2805	0.2853
Cantabria	0.3153	0.3306	0.2863	0.3293	0.2708	0.2695	0.2614	0.2754
Castilla-La Mancha	0.3009	0.3292	0.2708	0.3120	0.2691	0.2599	0.2775	0.2720
Castilla y Leó	0.3246	0.3418	0.2976	0.3375	0.2927	0.2842	0.2969	0.2966
Catalunya	0.3233	0.3236	0.3087	0.3321	0.2883	0.2902	0.2746	0.2928
Comunidad Valenciana	0.3173	0.3244	0.2980	0.3268	0.2739	0.2757	0.2633	0.2771
Extremadura	0.3170	0.3499	0.2863	0.3250	0.2716	0.2678	0.2750	0.2725
Galicia	0.3043	0.3230	0.2757	0.3139	0.2804	0.2834	0.2677	0.2841
La Rioja	0.3066	0.3228	0.2792	0.3177	0.2556	0.2523	0.2527	0.2588
Madrid	0.3357	0.3333	0.3270	0.3419	0.3031	0.3085	0.2888	0.3059
Navarra	0.2982	0.2997	0.2785	0.3117	0.249	0.2468	0.2346	0.2557
Euskadi	0.3053	0.3245	0.2768	0.3187	0.2706	0.2654	0.2669	0.2755
Murcia	0.3311	0.3535	0.3000	0.3431	0.2814	0.275	0.2825	0.2849
Ceuta	0.3279	0.3377	0.3161	0.3371	0.3069	0.3025	0.3124	0.3094
Melilla	0.3235	0.3446	0.3039	0.3285	0.297	0.2915	0.3027	0.2982
Spain	**0.3272**	**0.3353**	**0.3091**	**0.3357**	**0.2879**	**0.2779**	**0.2895**	**0.2910**

Source: own elaboration.

Table A2. Indices of regional concentration on the whole population (IG), on the group of women (IG_{Women}), men (IG_{Men}) and gender (IG_{Gender}) concentration index. All have been calculated from the variable annual gross earnings per hour.

Acknowledgements

The authors disclose receipt of the following financial support for the research, authorship and/or publication of this article: Grant of the Spanish Ministry of Economics and Competitiveness. CSO2013-43054-R and UTA MAYOR 2015. The research leading to these results has received support under the European Commission's 7th Framework Programme (FP7/2013-2017) under grant agreement n°312691, InGRID-Inclusive Growth Research Infrastructure Diffusion.

Author details

Beatriz Larraz[1*], Mónica Navarrete[2] and José Manuel Pavía[3]

*Address all correspondence to: beatriz.larraz@uclm.es

1 Statistics Department, University of Castilla-La Mancha, Toledo, Spain

2 College of Business Administration (Escuela Universitaria de Administración y Negocios) University of Tarapacá, Arica, Chile

3 University of Valencia, Applied Economics Department, Valencia, Spain

References

[1] Stiglitz J. E. The Price of Inequality: How Today's Divided Society Endangers Our Future. New York: W. W. Norton & Company; 2012.

[2] International Monetary Fund. Income Inequality and Fiscal Policy. International Monetary Fund Staff Discussion Note. 2012; SDN/12/08.

[3] Gini, C. Sulla misura della concentraciones e della variabilitá dei caratteri (On the measurement of concentraciones and variability of characters). Atti del R.Istituto Veneto di Science, Lettere ed Arti. 1914; 73: 1203–1248.

[4] International Monetary Fund. Causes and Consequences of Income Inequality: A Global Perspective. Era-Dabla Norris, Kalpana Kochhar, Nujin Suphaphiphat, Frantisek Ricka, Evridiki Tsounta. International Monetary Fund Staff Discussion Note. 2015; SDN/15/13.

[5] Emakunde. Sistema de Indicadores. Igualdad mujeres y hombres en la CAE (Indicators System. Equality for women and men in the Basque Country). Vitoria: Instituto Vasco de la Mujer; 2012.

[6] European Institute for Gender Equality (EIGE). European Union. 2015. Available from: http://eige.europa.eu/node/435. Updated Table, July 14, 2015. [Accessed: February 25, 2016].

[7] Larraz, B. Decomposing the Gini inequality index: an expanded solution with survey data applied to analyze gender income inequality. Sociological Methods & Research. 2015; 44(3): 508–533.

[8] Eurostat. NUTS 2013 classification (Nomenclature of territorial units for statistics). 2015. Available from: http://ec.europa.eu/eurostat/web/nuts/overview. [Accessed: March 08, 2016].

[9] Micro-data of the Structure of Earnings Survey, 2010. Spanish Statistics Institute. 2012.

[10] Methodology of the Structure of Earnings Survey, 2010. Spanish Statistics Institute. 2012.

[11] Eurostat. Gini coefficient, Source SILC. Income and Living Conditions Statistics, 2012. http://appsso.eurostat.ec.europa.eu/nui/show.do?dataset=ilc_di12&lang=en. Access date: November 20, 2012.

[12] Giorgi, G. M. Gini's scientific work: an evergreen. Metro-International Journal of Statistics. 2005; LXIII: 299–315.

[13] Basulto, J. and Busto, J.J. Gini's concentration ratio. Electronic Journal for History of Probability and Statistics. 2010; 6: 1–42.

[14] Gini, C. Variabilitá e Mutabilitá: contributo allo Studio delle distribuzioni e delle relazioni statistiche (Variability and mutability : contribution to the study of distributions and statistical reports). Rome, Italy: Universitá de Cagliari; 1912.

[15] Moran, P. Notes on continuous stochastic phenomena. Biometrika. 1950; 37:17–23.

[16] Moran, P. The interpretation of statistical maps. Journal of the Royal Statistical Society B. 1948; 10: 243–251.

[17] Geary, R. The contiguity ratio and statistical mapping. The Incorporated Statistician. 1954; 5(3): 115–127.

[18] Cliff, A. and Ord, J. Spatial Processes: Models & Applications. London: Pion; 1981.

[19] Wackernagel, H. Multivariate Geostatistics. An Introduction with Applications. 2nd Edition. Berlin: Springer Verlag; 1995.

[20] Emery, X. Geoestadística Lineal (Linear Geostatistics). Departamento de Ingeniería de Minas. Facultad de Ciencias Físicas y Matemáticas. Santiago de Chile: Universidad de Chile; 2000.

[21] Matheron, G. Random Functions and Their Applications in Geology. Geostatistics, a Colloquium. Plenum Press. 1970; 79–87.

[22] Cressie, N. and Hawkins, D.M. Robust estimation of the variogram. Journal of the International Association of Mathematical Geology. 1980; 12(2): 115–125.

[23] Montero, J. M. and Larraz, B. Introducción a la Geoestadística Lineal (Introduction to Linear Geostatistics). A Coruña: Netbiblo Ed; 2008.

Chapter 4

Spatial Optimization of Urban Cellular Automata Model

Khalid Al-Ahmadi, Mohammed Alahmadi and
Sabah Alahmadi

Additional information is available at the end of the chapter

http://dx.doi.org/10.5772/64788

Abstract

Although cellular automata (CA) offer a modelling framework and set of techniques for modelling the dynamic processes of urban growth, determining the optimal value of weights or parameters for elements or factors of urban CA models is challenging. This chapter demonstrates the implementation of a calibration module in a fuzzy cellular urban growth model (FCUGM) for optimizing the weights and parameters of an urban CA model using three types of algorithms: (i) genetic algorithm (GA), (ii) parallel simulated annealing (PSA) and (iii) expert knowledge (EK). It was found that the GA followed by EK produced better and more accurate and consistent results compared with PSA. This suggests that the GA was able to some extent to understand the urban growth process and the underlying relationship between input factors in a way similar to human experts. It also suggests that the two algorithms (GA and EK) have similar agreement about the efficiency of scenarios in terms of modelling urban growth. In contrast, the results of the PSA do not show results corresponding to those of the GA or EK. This suggests that the complexity of the urban process is beyond the algorithm's capability or could be due to being trapped in local optima. With this satisfactory calibration of the FCUGM for the urban growth of Riyadh city in Saudi Arabia by using CALIB-FCUGM, these calibrated parameters can be passed into the SIM-FCUGM to simulate the spatial patterns of urban growth of Riyadh.

Keywords: cellular automata, urban growth, calibration, genetic algorithm, parallel simulated annealing, Riyadh

1. Introduction

Linear, static, top-down, descriptive and explanatory models cannot adequately help to explain and reflect the essence of urban phenomena. With deeper understanding of urban phenomena, scientists have begun to recognize that cities are not uniform or a single type of phenomenon but more typically hierarchies of complex systems. As complexity theory and its properties have developed over the last three decades based on studies of non-linear systems, fractals, bifurcations, self-organization and chaos theory, cities have gradually become regarded as spatially complex systems [1–3]. A city can be characterized as a non-linear, open, complex, self-organizing and emergent system, which is far from being in equilibrium [1, 4, 5]. Urban growth dynamics are the direct consequence of the actions of individuals, public and private corporations (local agents) acting simultaneously over urban space and time. Therefore, cities are the spatial result over time of all these influences, which continuously contribute to shaping a city (aggregate global form). Cellular automata (CA) offer a modelling framework and set of techniques for modelling the dynamic processes and outcomes of such self-organizing systems [6]. CA techniques provide a way of simulating a self-organization process over geographical space and time [6, 7] and demonstrate significant potential benefits for urban modelling from the late 1980s due to their simplicity, flexibility and transparency [8–17]. However, Wu [18] argued that calibration of urban CA models is challenging when one seeks to determine the optimal value of weights or parameters for elements or factors of a model. If one can find optimal values, the results from running the model are likely to be greatly improved. With this in mind, the authors designed, implemented and evaluated a prototype for calibrating a stochastic, high-dimensional (up to 95) and non-linear urban CA model.

A fuzzy cellular automata model of urban growth was presented in Ref. [19]. Al-Ahmadi et al. presented an urban planning tool for the city of Riyadh, Saudi Arabia, which is one of the world's major cities undergoing rapid development. At the core of the system is a fuzzy cellular urban growth model (FCUGM), which is capable of simulating and predicting the complexities of urban growth. This model was shown to be capable of replicating the trends and characteristics of an urban environment during three periods: 1987–1997, 1997–2005 and 1987–2005. In another paper [20], the model was used to study and evaluate several different planning scenarios, both baseline ones and scenarios that relate to actual Saudi government policy. The results demonstrated that the model was capable of predicting plausible patterns of future urban growth. The model also has wider implications for use as a spatial planning support tool for urban planners and decision-makers in Saudi Arabia. A description of the application of fuzzy logic in the calibration of the FCUGM was presented in Ref. [21]. Along with calibration, one of the most significant aspects of any model is to verify, validate and assess its performance. The focus of the work published by Al-Ahmadi et al. [22] was on the techniques used to validate the performance of the FCUGM. They presented seven different validation metrics including visual inspection, accuracy and spatial statistics, metrics for spatial pattern and district structure detection as well as spatial multi-resolution validation.

The aim of this chapter is to describe the implementation of a calibration module in the FCUGM for optimizing the parameters for different modes and scenarios of the FCUGM using three types of algorithms: (i) genetic algorithm (GA), (ii) parallel simulated annealing (PSA) and (iii) expert knowledge (EK). These were applied over three periods [urban growth boundary

http://dx.doi.org/10.5772/64788

(UGB)] including UGB I (1987–1997), UGB II (1997–2005) and UGB I + II (1987–2005). The FCUGM is a hybrid CA model for research in urban planning and urban growth. It aims to explore and explain the complex spatial patterns of urban growth and to support the spatial urban planning through its two modules namely CALIB-FCUGM, the calibration model, and SIM-FCUGM, the simulation model, which can be used for prediction. Although the FCUGM is based upon fuzziness, it is designed to use stochastically constrained CA models.

2. Study area: geographic situation, physical environment and urbanization process

The Kingdom of Saudi Arabia is situated at the furthermost part of south-western Asia and occupies approximately four-fifths of the Arab Peninsula, covering a total area of 2.25 million km^2 of which about 40% are desert lands, and a population of 22,673,538 million according to 2004 census. The city of Riyadh is situated on the Najd Plateau in the central region of the Arabian Peninsula and surrounded to the east by high land ridges and to the west by the convergence of valleys forming Wadi Hanifah and Mount Tuwaiq. Riyadh is one of the fastest growing cities in the Middle East. The annual rate of population growth in Riyadh has reached an average of 8.1% by natural increase and immigration, and according to recent forecasts, the population is expected to increase to 10 million by 2020. In parallel with this dramatic increase in population, the spatial extent of Riyadh has grown from less than 1 km^2 in 1920 to over 1150 km^2 in 2004.

The urbanization process of Riyadh during the period between 1750 and 2004 has passed through four main phases of development namely the pioneer phase, the pre-establishment phase, the establishment phase and the oil-boom and post-oil boom phase. Broadly, the increase in wealth, building of the railway, the inauguration of the airport and transferring government agencies from Jeddah to Riyadh and the need to build new ministries and hundreds of houses has had a significant impact on the urban growth of Riyadh. This high rate of growth in population and areas has not been met with an adequate expansion of services, management capacity and development intervention. As a result, several types of problems have manifested, for example, the spread of slums and squatter settlements, a shortage of services for large parts of the city and a growth in demand for housing accompanied by land and transportation difficulties. After examining the main three Master Plans of Riyadh, the results indicate that most of the criticisms of the first and second Master Plans were based on the fact that they did not adequately anticipate the size of urban growth, which took place in Riyadh; this was because much of the development occurred beyond the boundaries designated by the plan. This resulted in unexpected urban sprawl. Another weakness aspect of these two Master Plans was that they were formulated on the basis of moderate economic growth rate. Consequently, they could not have anticipated the economic effect of oil boom in the 1970s and its adverse effect on the city's physical growth in terms of density and scale. This suggested the need for a tool to generate different scenarios of urban growth and test the potential physical and environmental impact for each scenario. Planning authorities, urban planners and decision-makers in Saudi Arabia have recently, however, begun to use spatial analytical and other planning tools to simulate and evaluate the consequences of urban planning policies prior to implementing them. Such tools can help to explore plans, policies and other factors underpinning and influencing processes of urban growth in the recent past, which can in turn lead to a better understanding the current factors influencing urban growth and ultimately in making more reliable predictions. Based on the use of software applications and tools, one can generate and evaluate the consequences of diverse future scenarios for urban growth by answering 'what if' type questions.

In this chapter, the term 'urban growth' refers to the physical transformation of vacant, dessert or agricultural land to urban land by planning and building infrastructure and industrial, residential, retail, educational and other buildings and social and recreational facilities.

3. Uncertainty and global sensitivity analysis of FCUGM

Although many studies [8–17] have investigated models of urban growth-based CA, little attention has been paid to examining the uncertainty and errors in urban CA models. It has been hypothesised that urban CA models are influenced by uncertainties that might be generated from various sources such as the complex interaction between input factors and parameters, specification and structure of the model and quality of input data [23]. The structure of CA models is not error-free; however, like other computer models, they are affected by errors owing to poor or partial human knowledge, complexity of the process being investigated and limitations of technology [23, 24]. The impact of neighbourhood size and type on model outcomes of a GIS-CA urban growth model was analysed by Kocabas and Dragicevic [24]. They applied univariate sensitivity analysis to study the variations in model outcomes by changing one parameter at a time while other parameters were kept constant. They found that the size and type of neighbourhood parameters have a significant influence on CA model output. The use of such a technique is considered as local sensitivity analysis. It is, however, time-consuming and cumbersome if more than two parameters are allowed to vary simultaneously. It is also deterministic and static. It cannot mimic the non-linear, stochastic and dynamic features, which typically exist in urban models. The error propagation in urban CA simulation was examined by Yeh and Li [23] through using a Monte Carlo Simulation (MCE). When MCE is applied, the spatial variables are perturbed so that the sensitivities of perturbations in urban simulation can be assessed in terms of errors in the outcome of simulation.

The FCUGM models the spatial pattern of urban growth using three modes: Mode 1, Mode 2 and Mode 3. The three modes differ in the structure of the fuzzy IF-THEN rule because different structures of transition rules might generate different simulation outcomes. The FCUGM can simulate spatial patterns of urban growth under nine scenarios [21]. An uncertainty and global sensitivity analysis (UGSA) was undertaken on all of the nine scenarios in the three modes of the FCUGM in order to assess the effects of uncertainties in the input variable (independent factor) and on the output variable (dependent factor). The advantage of using global rather than local sensitivity analysis is that the former is dynamic, stochastic and apportions the output uncertainty to the uncertainty in all of input variables. It evaluates the effect of one input variable while all of the others are varied as well. In contrast, the local perturbative approach is based on partial derivatives. The effect of the variation in one input factor is evaluated when all of the others are kept constant at their central value [25]. In the FCUGM, UGSA can provide an initial estimation of the quality of each mode and scenario in terms of understating urban growth of Riyadh. Since each mode and scenario has a different specification and structure, UGSA will be applied to help in identifying the most appropriate one.

The MCS technique was selected to undertake the UGSA because it has been applied successfully in a variety of applications including financial risk and statistical physics [26]. In addition, the MCS is one of the frequently applied techniques for computer simulations or numerical experiments. In terms of urban CA models, Yeh and Li [27] claimed that MCS tended to be most appropriate for the investigation of error propagation in urban CA simulation, particularly when mathematical models are difficult to define. Moreover, applying MCS has advantages since urban CA models cannot be modelled explicitly based on mathematical equations. Although one of the main drawbacks of MCS is the computation time required to generate a large number of samples, yet recent advancements in computer technology have reduced this problem [23]. MCS is relatively simple and straightforward to apply. It is generally based on generating numerous evaluations (runs) of the model with randomly selected input values for variables. For each trial or run, the input variables are assigned to random values based on selected input distributions and the value of each output variable recorded [25]. The results of MCS are, however, only an approximation (not exact) of the true value [26].

Scenario	Mean	SD	Skewness	Kurtosis	90% certainty value
Mode 1—Scenario 1	0.277	0.142	0.079	1.987	0.523
Mode 1—Scenario 2	0.505	0.15	0.154	2.053	0.656
Mode 1—Scenario 3	0.319	0.159	−0.187	1.899	0.701
Mode 1—Scenario 4	0.458	0.189	0.343	2.859	0.507
Mode 2—Scenario 1	0.508	0.109	−1.257	2.468	0.628
Mode 2—Scenario 2	0.324	0.164	−0.062	1.891	0.518
Mode 2—Scenario 3	0.458	0.189	0.343	2.859	0.674
Mode 2—Scenario 4	0.187	0.132	0.754	2.753	0.423
Mode 3—Scenario 1	0.223	0.149	0.572	2.538	0.315

Table 1. Output distribution statistics of the uncertainty analysis of FCUGM's scenarios using Monte Carlo simulation (MCS).

These were chosen to generate and evaluate different urban growth scenarios based on different planning objectives. In the context of the FCUGM, the independent variables are the parameter values of input variables while the dependent variable is the output mean square error (MSE) of the scenario. Thus, the UGSA will examine the effect of the variations in parameters values on the MSE outcome. There are no rules for selecting the 'best' number of iterations for performing UGSA primarily because it is problem-dependent. Sufficient iterations are essential, however, to determine statistically the relevant response distribution. Technically 1000 to 10,000 trials are usually good measures in terms of the number of trials [26]. The MCS was run 5000 times for each scenario. The uncertainty in the parameters of the input variable was represented by a uniform distribution with lower and upper bounds corresponding to each input variable. Each trial will be evaluated by calculating the MSE of the differences between the observed and simulated urban maps. Five distribution statistics were computed to assess the output variable (MSE) resulting from MCS for each scenario including: mean, standard deviation (SD), skewness, kurtosis and 90% certainty value (CV), as shown in **Table 1**. The skewness measures the extent to which the MSE values cluster to one side or the other of the mean. When most values and a higher number of occurrences cluster towards the left tail, this implies that they should provide a good solution. The kurtosis measures the sharpness of the distribution. A kurtosis greater than three indicates a high peak of occurrences, while less than three indicates a flat top [28]. The CV represents the value of the MSE that 90% of the outputs (trails) less than the returned CV. Thus, the lower the CV value, the better the scenario. **Figure 1A–I** shows the occurrences of MSE generated from each scenario; this indicates the empirical estimation of MSE for the random combinations of the input parameters.

As illustrated in **Table 1**, Mode 1—Scenario 4, Mode 2—Scenario 4 and Mode 3—Scenario 1 generated the best performance with the lowest certainty values of 0.507, 0.423 and 0.315, respectively. This means that 90% of the occurrences (iterations) have a MSE with 0.507, 0.423 and 0.315 for these three scenarios. In addition, as shown in **Figure 1**, these three scenarios present a similar pattern where most of the occurrences are clustered towards the left side, with a low MSE output and thus better performance. This is supported quantitatively by accounting for the higher skewness rates with 0.343, 0.754 and 0.572 for Mode 1—Scenario 4, Mode 2—Scenario 4 and Mode 3—Scenario 1, respectively. In contrast, Mode 1—Scenario 2, Mode 1—Scenario 3 and Mode 2—Scenario 1 show the highest certainty values of 0.656, 0.701 and 0.628, respectively. This indicates that 90% of the solutions are below a relatively high MSE range (0.65–0.701). Note that the structure of modes is based on the number of fuzzy variables embedded in each fuzzy rule and the structure of scenarios is founded on the number and type of urban growth factors, specifically the transportation support factor (TSF), urban agglomeration and attractiveness factor (UAAF) and topographical constraints factor (TCF) [21]. It can be inferred that the number of urban growth factors in each scenario has, to a large extent, considerable influence on the performance of the scenario. For example, the three scenarios that showed the best performance namely Mode 1—Scenario 4, Mode 2—Scenario 4 and Mode 3—Scenario 1, are the only scenarios among the total of nine that included the three urban growth factors TSF, UAAF and TCF.

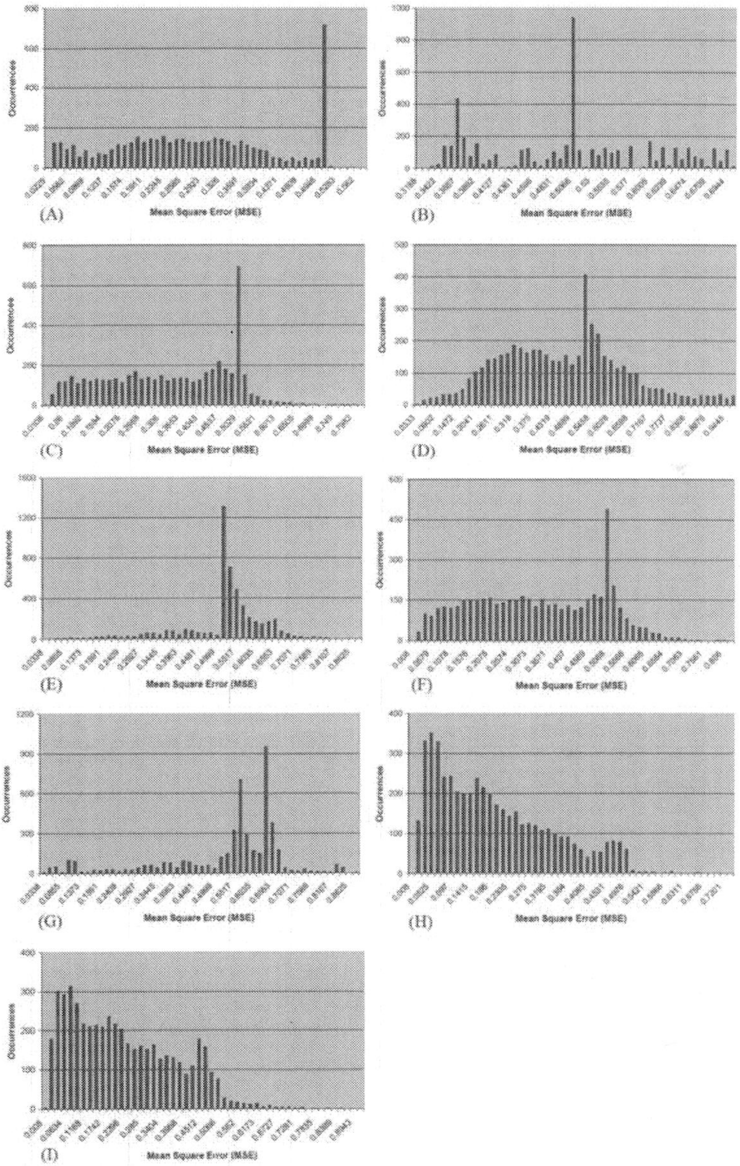

Figure 1. Uncertainty analysis for scenarios of the FCUGM using Monte Carlo simulation: (A) Mode 1—Scenario 1, (B) Mode 1—Scenario 2, (C) Mode 1—Scenario 3, (D) Mode 1—Scenario 4, (E) Mode 2—Scenario 1, (F) Mode 2—Scenario 2, (G) Mode 2—Scenario 3, (H) Mode 2—Scenario 4 and (I) Mode 3—Scenario 1.

In contrast, the remaining six scenarios embed only one or two. This suggests that the urban growth process in Riyadh can be modelled more accurately by integrating these three factors into a single scenario, rather than just using one or two of them. In addition, one can deduce that the higher the number of fuzzy variables embedded in each single fuzzy rule in the mode, the better the performance of that mode. Mode 3—Scenario 1, for example, embeds three fuzzy variables in each rule and accounts for the highest certainty value, that is, 90% of the 5000 evaluations produced a low MSE with less than 0.315, which indicates that such a mode structure is better than any other. When the number of fuzzy variables in the fuzzy rule decreases, the MSE decreases, for instance, Mode 2—Scenario 4 (two fuzzy variables with 0.423) and Mode 1—Scenario 4 (one fuzzy variable with 0.507). However, the high accuracy produced by Mode 3 involved a high computation time. One can see that as the fuzzy variables in the fuzzy rule increase, the simulation time increases exponentially. For example, the average computation time was 4.5, 8 and 19 hours for scenarios in Mode 1, Mode 2 and Mode 3, respectively.

With respect to the scenarios in Mode 1 (except for the best one, Scenario 4), it can be inferred that urban growth in Riyadh is influenced by transportation support (Scenario 1 with a CV of 0.523) more than socio-economic services (Scenario 2 with a CV of 0.656) and topographical constraint factors (Scenario 3 with a CV of 0.701). With regard to the scenarios in Mode 2 (with the exception of Scenario 4), it can be inferred that the process of urban expansion in Riyadh city is moderately affected by integrating the transportation support with socio-economic services (Scenario 2 with 0.518 as CV) more than by integrating transportation support with topographical constraint factors (scenario 1 with a CV of 0.628) or socio-economic services with topographical constraints factors (Scenario 3 with a CV of 0.674).

4. Calibration of the FCUGM

The calibration process of the FCUGM is undertaken by a module called the CALIB-FCUGM. This consists of several interlinked sub-models that are processed sequentially either once or several times during a calibration period. The CALIB-FCUGM aims to provide the SIM-FCUGM, the module by which the simulation is executed, with the optimal parameter values or weights of spatial variables to enable realistic generation of urban patterns. The CALIB-FCUGM optimizes parameters by three different algorithms, namely GA, PSA and EK.

4.1. Basic process flow of CALIB-FCUGM

The stages of the CALIB-FCUGM are illustrated in **Figure 2**. The main procedures of the CALIB-FCUGM fall into four stages: (i) Input Variables Weighter, Fuzzy Distance Decay Quantifier, Fuzzy Input Variables Integrator and Fuzzy Input Variables Normalizer (yellow boxes); (ii) Fuzzy model (green boxes); (iii) CA model (blue boxes); and (iv) Optimization Algorithms (grey box), as shown in **Figure 2**. Most boxes in **Figure 2** are a sub-model of CALIB-FCUGM; it takes some outputs from the preceding sub-model and feeds the subsequent sub-model with some inputs. The dashed boxes indicate that this sub-model includes parameters, which require to be optimized. The calibration process works sequentially. It begins by reading input variables into the Input Variables Weighter, by which a weight is assigned to each input variable reflecting its corresponding importance to other variables. Next, the weighted input variables are passed into the Fuzzy Distance Decay Quantifier to compute the effect of the distance decay of each variable by optimizing the distance decay parameters. These weighted fuzzy variables are then fed into the Fuzzy Input Variables Integrator, which integrates these weighted fuzzy variables into three fuzzy driving forces [19, 21]. These in turn are normalized

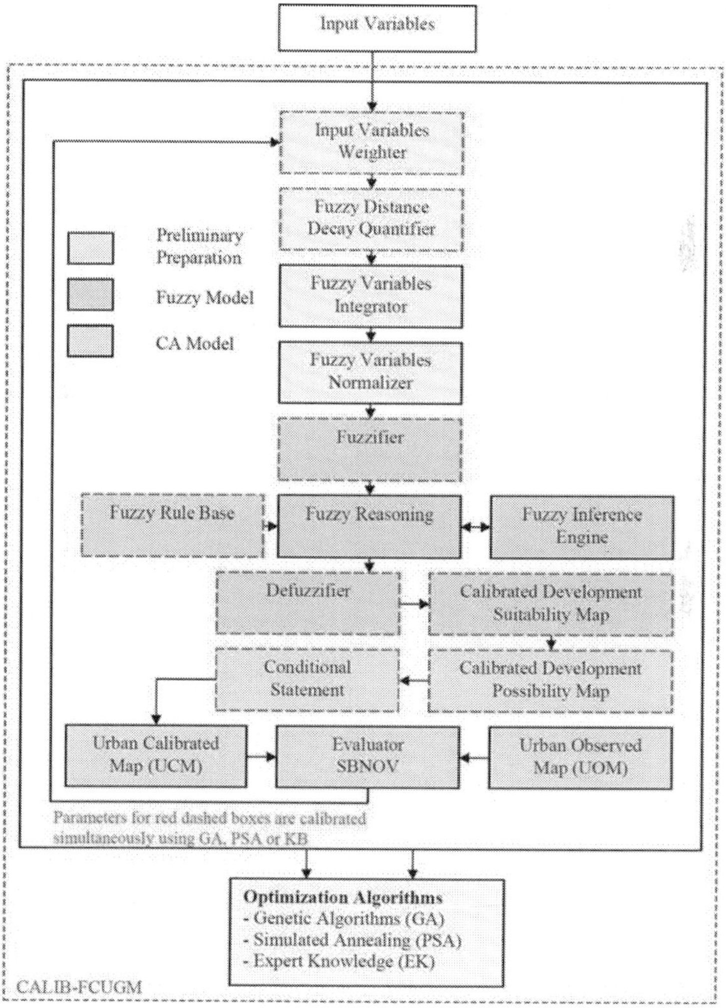

Figure 2. Calibration Process of the CALIB-FCUGM.

to between 1 and 100. The next stage involves passing these three fuzzy input variables into the fuzzy model and creating a 'calibrated development suitability map'. After adding a stochastic disturbance factor into the development suitability map, this map is called the 'calibrated development possibility map'. This calibrated development possibility map is then entered a conditional statement to decide whether a certain location can be considered as 'urban' or 'non-urban' based on both its development possibility and the calibrated transition threshold. The conditional statement outputs the final 'urban calibrated map (UCM)', which is a binary map (1 for urban and 0 for non-urban). Finally, the 'urban calibrated map' is read by the Evaluator, in order to assess the accuracy of this map, and compared with the 'urban observed map (UOM)' (which is also a binary map), by computing the error between the two maps by calculating the best net objective value (BNOV). All of this procedure is generated several times according to the characteristics of each of the three algorithms GA, PSA and KB. Note that the CALIB-FCUGM module works automatically after a user enters the input variables to the module. The outcome of the CALIB-FCUGM module is an optimal set of parameters and weights. This will be read into the SIM-FCUGM module to simulate urban development. As the FCUGM is a loosely coupled model, the output of the CALIB-FCUGM is read by the SIM-FCUGM by manual entry.

4.2. Feasible solution of the CALIB-FCUGM

As stated earlier, the main aim of the CALIB-FCUGM is to find the optimal set of weights and parameters for each scenario of the FCUGM. Each candidate solution provided by the CALIB-FCUGM is a set of weights or parameters, which vary according to their associated range (predefined upper and lower bounds). **Table 2** shows the total number of weights and parameters, which are calibrated for each scenario. As shown, the number of weights and parameters for scenarios is different. This is due to the difference in the number of fuzzy variables employed in each scenario. This affects the number of input variables, number of weights, number of distance decay parameters and other parameters because all these parameters are used to build fuzzy variables.

Modes and scenarios	Number of weights and parameters
Mode 1—Scenario 1	57
Mode 1—Scenario 2	59
Mode 1—Scenario 3	57
Mode 1—Scenario 4	65
Mode 2—Scenario 1	63
Mode 2—Scenario 2	69
Mode 2—Scenario 3	69
Mode 2—Scenario 4	93
Mode 3—Scenario 1	99

Table 2. Number of weights and parameters for scenarios in the FCUGM.

4.3. Objective function of CALIB-FCUGM

The performance of the GA, PSA or EK algorithms is evaluated based on the quality of the final solution acquired by the algorithm. In relation to the quality of the final solution, the value of the objective function (cost function), which is also referred to as an fitness function in GA and energy function in PSA, is the major criterion for assessing performance of the algorithm. The effectiveness of any iterative algorithm such as GA or PSA depends heavily on having an efficient objective function. The purpose of the objective function is to determine for any given configuration of the search space a value that represents the relative accuracy of that configuration or solution. In the CALIB-FCUGM context, the robustness of the solution can be considered as an error and the objective function aims to minimize the error between the UOM and the UCM.

There are several techniques for measuring errors, which can be used in the FCUGM problem such as total absolute error (TAE), mean absolute error (MAE), MSE, root mean square error (RMSE), normalized root mean squared error (NRMSE), relative operating characteristic (ROC), confusion matrix (CM) and Kappa Index of Agreement (KIA). The measurement of differences in errors between the observed and simulated images has been performed in different ways by various authors. A CM was used by Wu and Webster [29] to evaluate the accuracy of the simulated image against the observed one. The MSE and the MAE were used by Li and Yeh [30, 31] for measuring errors between simulated and observed images in a study involving modelling urban developments. The MSE also used by Kim [32] for measuring the accuracy between the observed and probability images as a way of validating results from calibration process. The NRMSE was used by Heppenstall [33] as a fitness function to validate the calibration results of a GA and to measure the error between the observed and predicted spatial multi-agent model for petrol prices. In addition, Pontius and Schneider [34] applied and explained how to use the ROC technique to examine how well a probability map portrays the likely locations of a category of new development. The Leica ERDAS image processing application uses RMSE for measuring the error of image rectification and KIA for validating image classification results.

As a result, in the FCUGM, the authors selected two types of measures, one to verify the calibration results and the other for testing the simulating results. Although most of the techniques are appropriate for verifying the performance of simulation processes, few of them are suitable for doing this for calibration. This is because the calibration process in the FCUGM requires the candidate solution to be assessed in each iteration, while in the simulation process, the results are verified once at the end. Consequently, the MSE and RMSE were selected to validate the results of the CALIB-FCUGM for several reasons. First, they are the most well known and widely used techniques of error measurement [35]. Second, they are efficient for validating the performance in a cell-by-cell manner, which is the case in calibrating the FCUGM, and they will be calculated in this research as given below in Eqs. 1 and 2:

$$\mathrm{OFI} = \mathrm{MSE} = \frac{\sum_{i=1}^{n}\left(O_{ij} - C_{ij}\right)^{2}}{n} \tag{1}$$

$$\text{OFI} = \text{RMSE} = \sqrt{\frac{\sum_{i=1}^{n}\left(O_{ij} - C_{ij}\right)^2}{n}} \tag{2}$$

where OFI is the objective function (MSE or RMSE) of a location *ij*; O_{ij} is the urban observed state at location *ij*; C_{ij} is the urban calibrated state at location *ij*; and *n* is the number of locations or cells.

Although several research studies have only applied a straightforward objective or fitness function such as the MSE or RMSE as measure for error, little attention has been paid to measuring the effect of constraints. It has been claimed that GA and PSA are stochastic algorithms and have to be constrained to explore only the search space with desired values. The author argues, however, that it would be much better to compute the overall net objective value (NOV) as well, because such a measure includes a weighting system with objective functions and implemented constraints through penalty functions, which add to the overall objective value. The net objective value, therefore, is penalized as the set of design variables moves further out of bounds or does not meet a constant constraint value. The NOV can be computed as shown in Eq. 3:

$$\text{NOV} = \left(OF_i \times W_{OF_i}\right) + \left(PF_i \times W_{PF_i}\right) \tag{3}$$

where OF_i is the objective function (MSE or RMSE) for a solution *i*; W_{OF_i} is the weight of the objective function (MSE or RMSE) for a solution *i*; PF_i is the penalty function for a solution *i*; and W_{PF_i} is the weight of the penalty function for a solution *i*.

It can be difficult, however, to compare NOV values from different experiments if the range and mean of the NOV are different in each case. Thus, to avoid this problem, the standardized net objective value (SNOV) will be used as shown in Eq. 4:

$$\text{SNOV}_i = \frac{\left(OF_i \times W_{OF_i}\right) + \left(PF_i \times W_{PF_i}\right)}{\text{Range}_i} \tag{4}$$

The penalty functions that will be used in the CALIB-FCUGM include two types of constraints: (i) equality (some calibrated parameter values have to be equal a constraint value) and (ii) inequality (some calibrated parameter values have to be less or greater than constraints). An example of the equality constraint is that the total calibrated weights should be equal to 100; if it is more or less than 100, the net objective value is penalized by adding this difference to the net objective value resulting in poorer solutions.

http://dx.doi.org/10.5772/64788

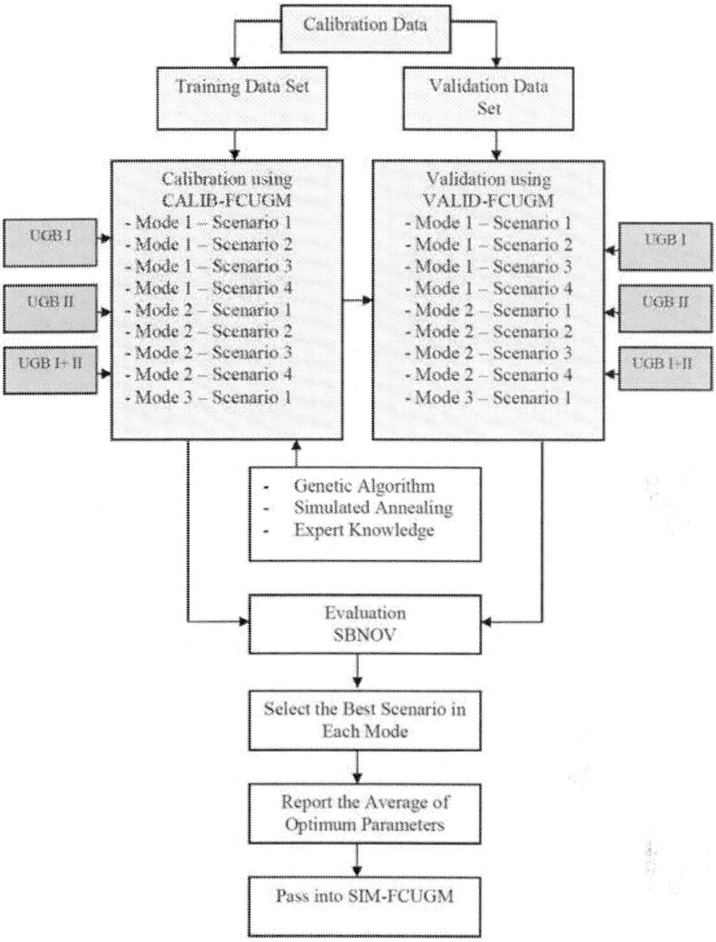

Figure 3. Steps of experiments for calibration process.

4.4. Experimental design of calibration process

In order to calibrate the FCUGM for acquiring the best set of parameters to generate a realistic simulation, several experiments were conducted. The experiments have eight aspects: (i) sample data set; (ii) calibration algorithms; (iii) mode; (iv) scenarios; (v) urban growth periods; (vi) training process; (vii) cross-validation process and (viii) calibration time. **Figure 3** illustrates the process of the calibration experiments. The best sample size for calibration is specified. Then, this data set is divided equally into two parts, one called 'training data set' and the other 'cross-validation data set'. The purpose of the former is to train the performance of the scenario of interest, while the latter aims to verify the calibrated parameters, which were generated by using the training data. The authors propose to calibrate the FCUGM urban growth process under three modes, which are comprised of nine scenarios as previously explained.

Each scenario is calibrated over three periods UGBI, UGBII and UGBI+II. The process starts by passing the training data sets into the CALIB-FCUGM, so the model is calibrated by three algorithms: GA, PSA and KB. Each scenario is calibrated five times for each period. Then, the parameters of the best solution are passed into the VALID-FCUGM, where the cross-validation data set exists, to verify the calibration results. The VALID-FCUGM is a static model, which validates the parameters as an off-line model. This process is conducted for each scenario over the three periods. Afterwards, the performance of the scenarios in terms of training and validation are evaluated and the best scenario in each mode is selected. Next, the mean of the optimal parameters for the best scenarios from applying the three algorithms is reported and passed into the SIM-FCUGM for simulation purposes.

4.5. Calibration data set

In terms of the calibration data set, it might be not appropriate to use the whole study area as a training data set because the volume of data is very large and could require very high levels of computational resources, which eventually affect the efficiency of the model. Moreover, spatial data are often not independent: the value of one observation is likely to be influenced by the value of another observation, so using the whole data set leads to the common problem of spatial autocorrelation (or spatial dependence) because values of variables at one location are more likely to be significantly associated with values at nearby locations. The high spatial dependency of variables is more likely to affect the accuracy of analysis and might lead to misinterpretation of the results. Random sampling is, however, a conventional way to overcome this problem [28, 31, 36].

The authors could not find any rules in the scientific literature about the 'best' type and size of random samples for calibrating urban models. An urban CA model was calibrated by Li and Yeh [30] using artificial neural networks by training the model using a proportional stratified random sampling method with a total of 3000 cells. The samples were proportionally randomly selected from different land use types, 50% (1500 points) being used as a training data set while the rest was used as test data set to verify the training results. In another study, Li and Yeh [31] calibrated the same model but with binary urban states (urban and non-urban) by applying the same spatial sampling method but with a total of 1000 samples, 50% for training and the remainder for validating training results. This suggests that the sample sizes reduce as the number of urban states decrease. There are many types of spatial sampling methods such as random, systematic, proportional random stratified, disproportional random stratified and clusters. It has been argued, however, that the stratified is better than the random sample, because the latter might supply redundant observations when sample locations are nearby to one another [36] and may exclude some smaller urban categories [31]. In any event, systematic sampling is not appropriate for the FCUGM problem because the urban and non-urban locations are randomly located and not systematically distributed. A random sampling method was used because it depends on the variable that is being investigated rather than the size of the variable area. In the case of the FCUGM, the urban state is the state by which the urban growth process and pattern are represented and measured. Thus, particular attention needs to be focused on the locations of the urban state rather than non-urban ones. In this sense, urban state locations need a more detailed monitoring or over-sampling, while maintaining adequate coverage of the non-urban portion of the sampled area. As a result, the proposed random sampling offers more intensity of samples for urban state locations with 60% of the total samples while 40% for the non-urban locations. With respect to the size of sample, Rogreson [36] claimed that the size of sample should be based on the accuracy that one seeks for estimation. Generally, the larger size of samples, the more accurate the estimation of means and proportions. Rogreson [36] claimed that, in general, accurate estimates can generally be obtained by choosing sample size according to Eq. 5.

$$n = \frac{Z^2}{4W^2} \tag{5}$$

where n is the size of sample; Z is the confidence intervals, that is, ±1.96 for 95% confidence interval and W is the width of the confidence interval.

Using Eq. 5, the total sample size for calibration in the FCUGM, with a 95% confidence interval and width within ±0.02, is ≈9600 samples. Fifty per cent of the total sample data set is randomly selected and used for training the calibration model, while the rest is used to verify the results of training, that is, 4800 cells were used for calibrating the model and 4800 cells were used for verifying the results of training.

5. The process of optimising algorithms within the CALIB-FCUGM

The basic theoretical foundation of GA and SA can be found in Refs. [37–40]. This section, however, examines these algorithms in relation to finding an optimal solution from the huge, non-linear and non-differential solution space of the FCUGM.

5.1. Genetic algorithms

In relation to GA, **Figure 4** shows how the GA works within the CALIB-FCUGM. Prior to starting the GA simulation, however, several decisions (FCUGM and GA parameters) have to be made as shown in **Figure 5**. After selecting suitable GA parameters, the GA simulation starts by generating an initial random population of a pre-specified number of chromosomes. Each chromosome is a solution out of all of the total potential solutions and is made of a number of genes. Each gene represents one parameter or weight value, which requires calibration. The gene is represented by a number of bits. Given that, **Table 3** displays the urban development scenarios in the FCUGM and the number of their genetic characteristics.

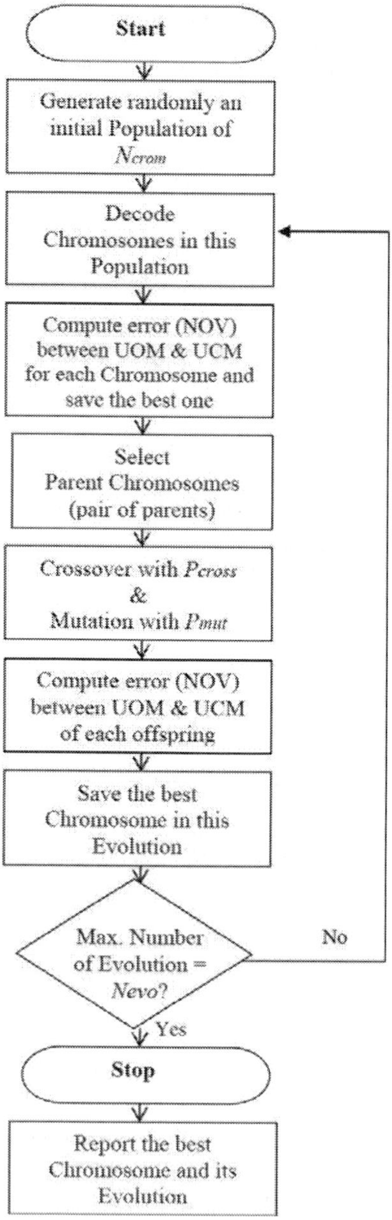

Figure 4. Flowchart of the Genetic Algorithm (GA) Process in the CALIB-FCUGM.

http://dx.doi.org/10.5772/64788

FCUGM Problem Parameters

- Define a space of feasible solution *S*
- Define an objective function, MSE and RMSE, *F*(*S*)
- Define a penalty function, Constraints, *F*(*C*)
- Define the net objective value *NOV*, *F* (*F*(*S*), *F*(*C*))

Genetic Algorithm Parameters

- Specify the size of population (no. of chromosomes) *Ncrom*
- Specify the maximum number of evolutions *Nevo*
- Define the selection method
- Define the crossover method and probability *Pcross*
- Define the mutation method and probability *Pmut*

Figure 5. The Parameters of the FCUGM and Genetic Algorithms (GA).

Modes and scenarios	Number of genes in each chromosome	Number of bits in each chromosome
Mode 1—Scenario 1	57	570
Mode 1—Scenario 2	59	590
Mode 1—Scenario 3	57	570
Mode 1—Scenario 4	65	650
Mode 2—Scenario 1	63	630
Mode 2—Scenario 2	69	690
Mode 2—Scenario 3	69	690
Mode 2—Scenario 4	93	930
Mode 3—Scenario 1	99	990

Table 3. Number of genetic characteristics in FCUGM scenarios.

The GA simulation starts by generating an initial random population (set of solutions) of a pre-specified number of chromosomes. Subsequently, each chromosome (solution) is decoded from bits into a certain value, that is, each parameter or weight is given a number within its bound. This is followed by evaluating the fitness of each individual solution in the initial population by calculating the error (according to Eqs. 1–4) between the UOM, the UCM and reported the NOV. Then, the best solution (lowest NOV value, i.e., BNOV) in this initial population is saved.

To create possible solutions for the next evolution (next population), three types of operators are applied including selection, crossover and mutation. These operators are described in more detail below. By the selection operator, two solutions are randomly selected proportion to their fitness values (based on the probabilistic function of fitness). The lower the NOV value, the more times it is likely to be selected to reproduce in the next generation. Next, the crossover procedure based on the crossover rate combines two solutions from the current evolution to produce two new solutions (offspring or children) for possible insertion in the next evolution. The mutation rules modify the solution by randomly altering one or more of the values of parameters or weights based on the mutation rate. Then, the best solution (lowest NOV value) in this evolution is saved. This iterative process continues until the maximum number of evolutions is performed (termination rule). CALIB-FCUGM checks whether or not the desired number of evolutions are met (termination rule), if not the population of the first evolution will be decoded and the same iterative processes continue. If the desired number of evolutions is met, then the CALIB-FCUGM will stop and evaluate the best solution in each evolution and select the best one and report the results of this solution in a form. The implication of selecting different GA parameters was examined by undertaking empirical experiments on different values of the parameters. **Table 4** shows the best control parameters of GA for FCUGM problem, which will be used for all subsequent experiments in this research.

GA parameters	Best options
Population size	Small (50)
Selection method	Tournament
Crossover probability	Medium (0.7)
Crossover method	Single point
Mutation probability	High (0.2)

Table 4. The best parameters of GA for the FCUGM problem.

5.2. Parallel simulated annealing

Similar to GA, prior to starting the PSA simulation, several decisions should be made as shown in **Figure 6**. It is worth noting that the PSA differs from the conventional SA in that sets of points (solutions) are run simultaneously in each control parameter rather than one single solution. **Figure 7** shows how the PSA works within the CALIB-FCUGM. The PSA simulation within the CALIB-FCUGM starts at a high temperature (control parameter) by generating a number of initial random solutions (Points) of the feasible solutions, each solution denoted as S0. Then, the error between the UOM and the UCM is measured by computing the NOV, the resultant value is denoted as NOV(S0), The lower the value of NOV(S0), the better the solution S0. The objective value NOV(S0) works to minimize the error (MSE, RMSE and meet its constraints).

http://dx.doi.org/10.5772/64788

FCUGM Problem Parameters

- Define a space of feasible solution *S*
- Define a neighborhood structure *NS*
- Define an objective function, MSE and RMSE, *F*(*S*)
- Define a penalty function, Constraints, *F*(*C*)
- Define an overall net objective value *NOV* (*F*(*S*), *F*(*C*))

Parallel Simulated Annealing Parameters

- Specify the initial control parameter *Ti*
- Specify the cooling function and cooling rate ∞
- Specify numbers of successful moves per control parameter steps *SM*
- Specify a termination criterion *Tf*

Figure 6. The parameters of the FCUGM and Parallel Simulated Annealing (PSA).

After calculating the NOV(S0), a small change in the initial solution S0 is brought about using a perturbation mechanism by which two weights or two parameters are randomly selected and their values are exchanged between them. This yields a new solution denoted as S1. Subsequently, a new cost function NOV(S1) is calculated in the same way as NOV(S0). Then, the results of the two objective functions NOV(S0) and NOV(S1) are evaluated. Whether the new solution is accepted or not is based on the following conditions:

If the NOV(S1) < NOV(S0), the objective function has declined (the error decreased) and the new solution S1 is accepted, and the current solution S0 is replaced with new solution, therefore, S0 is set to S1 and S0 = S1.

If the NOV(S1) > NOV(S0), the objective function has raised (the error increased) and is subjected to the metropolis criterion that will accept the new solution S1 according to the probability calculated as, exp((NOV(S0) – NOV(S1))/Ti), and the computed probability is compared to a uniformly distributed random number, R, between 0.0 and 1.0.

If R ≤ exp((NOV(S0) – NOV(S1))/Ti), the new solution is accepted, and the initial solution is replaced with new solution.

If R > exp((NOV(S0) – NOV(S1))/Ti), the new solution is rejected, and the initial solution stays in the same current state.

The preceding process is regarded as an iteration in SA algorithm. This process is repeated until the predefined number of successful moves (SM) in this particular temperature step is met. If the number of SM is met, it implies that a quasi-equilibrium state is reached at this particular control parameter step N and is liable to be reduced by the cooling function and cooling rate ∞ that were predefined. The processes will continue for a new control parameter step $N + 1$ unless the termination rule is met, i.e., the final control parameter Tf = 0.1. At this control parameter value, the algorithm will stop and provide the global optimal solution. The implication of selecting different PSA parameters was examined by undertaking empirical experiments on different values of the parameters. **Table 5** shows the best control parameters of PSA for FCUGM problem, which will be used for all subsequent experiments in this research.

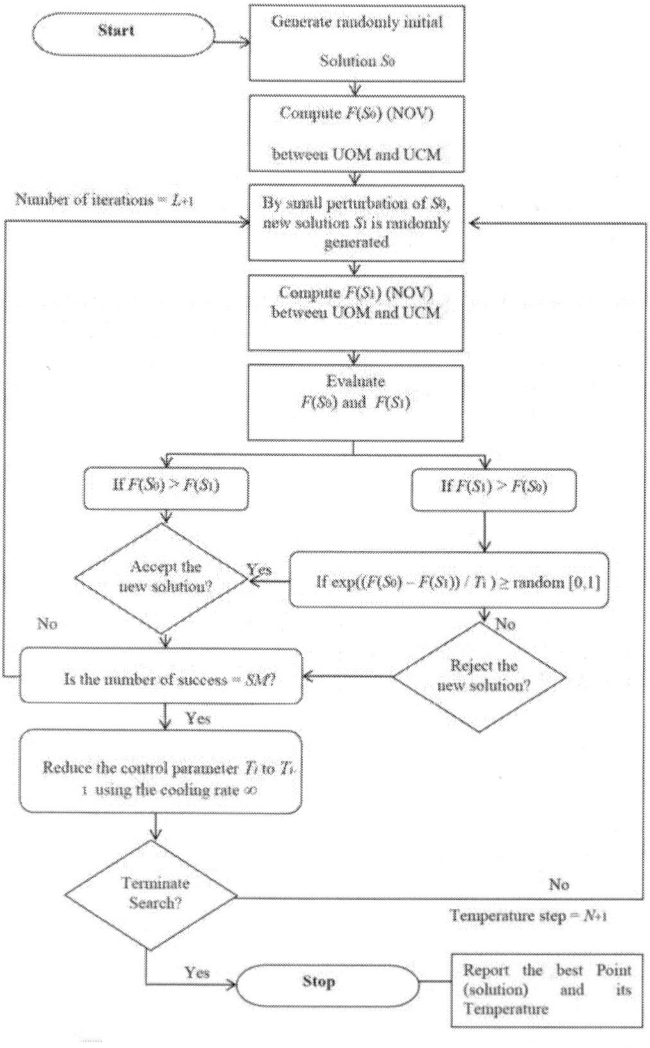

Figure 7. Flowchart of the PSA Process in the CALIB-FCUGM.

PSA parameters	Best parameters options and its values
Initial temperature	Moderate temperature—MT (60%)
Cooling function	Exponential cooling function—ECF
Cooling rate	Slow cooling—MC (0.9)
Number of successful moves	Medium number of successes—MNS (60)

Table 5. The best parameters of SA for the FCUGM problem.

5.3. Expert knowledge

In contrast to GA and PSA, by using EK the proper parameters and weights for the urban model are derived intuitively and empirically rather than automatically. In relation to the urban CA models, most studies calibrate parameters using a trial and error approach that combines the experience of the analyst. For example, in Ref. [29], the weights of urban factors and urban agglomeration are calibrated based on the analyst's views. The effect of distance decay parameters is calibrated empirically by Cheng and Masser [41] and Ward et al. [42]. In the FCUGM, the EK approach is not entirely based on the analyst's perspective. The parameters are calibrated on the foundation of the spatial structural analysis as well as the urban planner's experience. Thus, the calibration is not wholly qualitative in relying on a planner's view because quantitative results from the initial spatial structural analysis are used. Even so, the large number of parameters in some scenarios makes it very difficult for an expert to derive the proper parameter values.

6. Results and discussion

In this section, the FCUGM is calibrated using real data and the meaning of the calibrated values and the consistency of the calibration results, training and accuracy of the validation are discussed. In order to investigate the characteristics and features of the urban growth factors that might generate and affect the urban growth pattern of Riyadh city over the last 18 years, this period was divided into two intervals, namely UGB I and UGB II. The former represents the urban growth between 1987 and 1997, while the latter between 1997 and 2005. This division is not arbitrary; it is approximately the two intervals stated in the Government resolution on Urban Growth Boundary Policy. By calibrating the FCUGM over these two intervals, the authors would be able to assess the results and compare growth trends. The authors argue that combining the two periods (UGBI and UGBII) into one period (UGBI+II), which represents the urban growth between 1987 and 2005, so one can calibrate the model over the 18 years in one time, might provide an insight into changes in urban growth patterns. In this sense, the FCUGM was calibrated for three periods UGBI, UGBII and UGBI+II. The calibration process was carried out on nine different scenarios for each period, which are based on different urban growth factors and different transition rules. Thus, one can examine what are the best scenarios over each period and to what extent they correspond to the best scenarios over other periods.

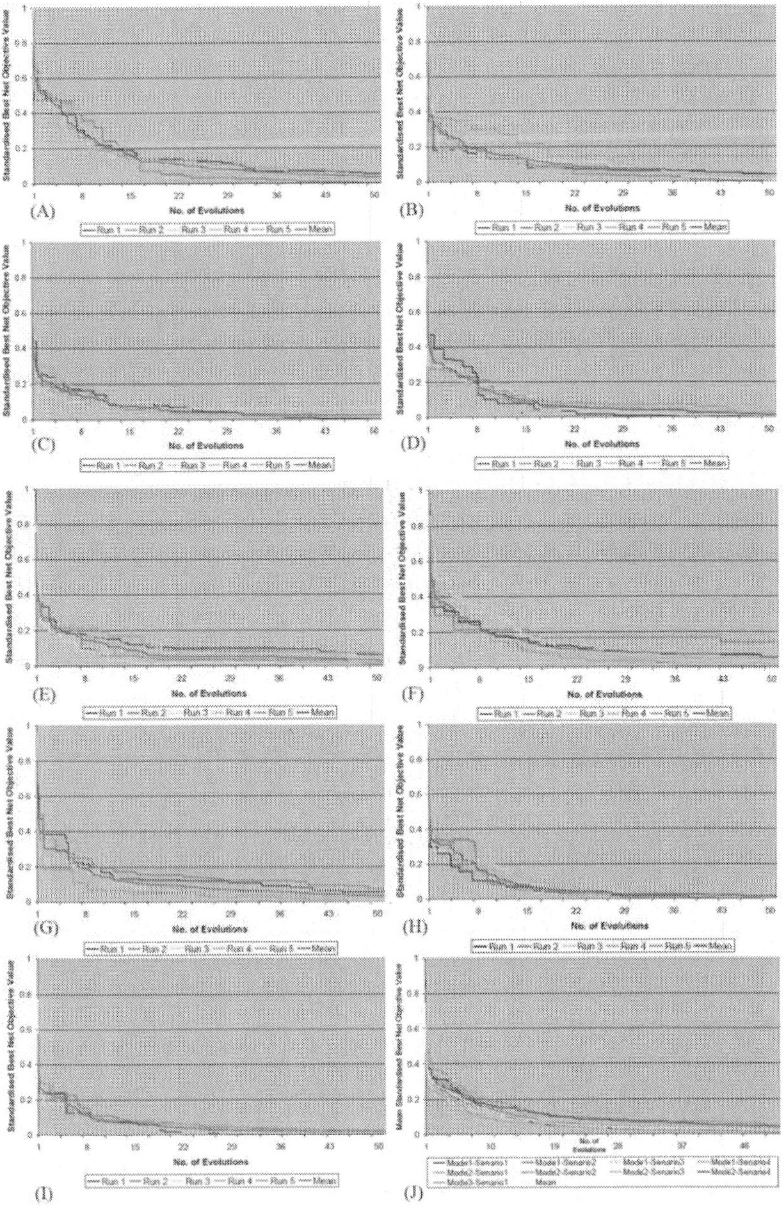

Figure 8. A–J: SBNOV evolution curves for (A) M1−S1, (B) M1−S2, (C) M1−S3, (D) M1−S4, (E) M2−S1, (F) M2−S2, (G) M2−S3, (H) M2−S4, (I) M3−S1 and (J) overall mean using GA, respectively.

As mentioned previously, the CALIB-FCUGM produced figures that show the progress of evolution and temperature for the GA and PSA respectively. **Figures 8A–J** and **9A–J** show 90 least-so-far standardized best net objective value (SBNOV) curves, five for each scenario and a mean SBNOV for each scenario using GA and PSA as a result of calibration FCUGM over the period UGB I + II. In terms of the progressive patterns, **Figures 8A–J** and **9A–J** show that the curves are concave, decreasing as the evolution increases in the GA and temperature decreases in PSA, i.e., the SBNOV declines as the evolution and temperature progress. Nevertheless, the degree of decrease and the values of starting and ending of SBNOV are varied from run to run, from one scenario to another and across all of the algorithms. Some curves decrease steeply in the early stages of evolution or temperature, while others decrease constantly in the middle or late stages. The variation in starting points of the GA might be attributed to dissimilar genetic characteristics in the different starting chromosomes. In the PSA, it might be because of the initial random states at different starting points. Broadly, convergence into the global solution (lowest SBNOV) decelerates as the evolution and temperature progress. The variation in ending points (the ends of curves' tails) of the GA and PSA might be because most runs converge to a narrow extent but generally do not converge altogether. This suggests that some performed better than others did. Some were possibly trapped in local minima.

In relation to the progress of the GA's evolution against the SBNOV, it can be seen that the SBNOV of the GA decreases in a consistent manner. For example, the SBNOV for most scenarios decreases exponentially with different degrees and little noise, indicating that errors are apparently decreasing as evolution progresses. This suggests the elitism feature of the GA, by which the best chromosome (solution) survives (passes) into the next evolution without any change, is working well. In contrast, the PSA shows considerable variations in the reduction of SBNOV against the PSA's temperature in different scenarios and modes. One possible reason for this variation is that the computation became stuck in local minima as shown in most scenarios. This is evident in the case of Mode 1—Scenario 2, Mode 1—Scenario 3 and Mode 1—Scenario 4, where the value of SBNOV decreases sharply in the early high temperature (first quarter) but afterwards (over the last three quarters) there was little or even no reduction of SBNOV. With respect to the convergence to the best global solution, it can be seen that most of the scenarios in the GA converged into very low SBNOV, broadly below 0.1, indicating positive performance of the algorithms across most scenarios. The higher convergence to the global solution are presented by Mode 1—Scenario 4, Mode 2—Scenario 4 and Mode 3—Scenario 1, where most evolution curves converge to a very narrow range towards curves' tails. This supports the argument generated as a result of the uncertainty and sensitivity analysis discussed above, that these three scenarios produced the higher certainty values. It also suggests that the structure of these scenarios and the urban growth factors embedded in them are most appropriate for understanding urban growth processes. The convergence to the best global solution in PSA was, however, varied without any apparent pattern of convergence. The variations were not only evident by scenarios but also by running within a single scenario. For example, Mode 1—Scenario 4 converges to a different solution with different SBNOV in each run, where the SBNOV ranges between 0.1 and 0.5. Thus, it can be deduced that PSA yielded poor solutions with inconsistent convergence in the global solution. However, only

Mode 2—Scenario 3 and Mode 2—Scenario 4 showed better convergence into low SBNOV for most of their scenarios.

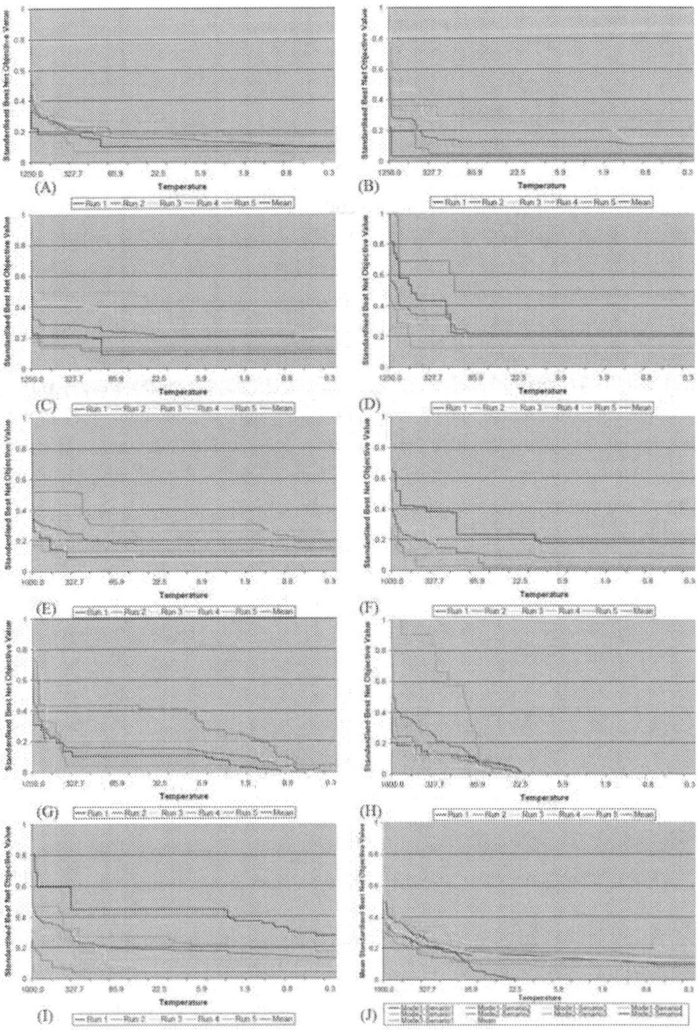

Figure 9. A-J: SBNOV evolution curves for (A) M1—S1, (B) M1—S2, (C) M1—S3, (D) M1—S4, (E) M2—S1, (F) M2—S2, (G) M2—S3, (H) M2—S4, (I) M3—S1 and (J) overall mean using PSA, respectively.

Figure 10A–F shows a comparison of the mean of Standardized Best Net Objective Value (SBNOV) in terms of all runs, training and validation of the optimum solution found by running CALIB-FCUGM five times for each scenarios using GA, PSA and EK. It can be seen from **Figure 8A–F** that, while there is some variation, there is broad correspondence in the performance of calibration between the algorithms, in terms of overall accuracy and validation. In terms of algorithm, the GA broadly produces highly consistent results with relatively low variations among different runs for each scenario. It can easily be observed that Scenario 4 in Mode 1, Scenario 4 in Mode 2 and Scenario 1 in Mode 3 account for by the lowest SBNOV generated from different runs. Scenario 2 in Mode 1 and Scenario 3 in Mode 2 yield the worse solution with high SBNOV in most runs.

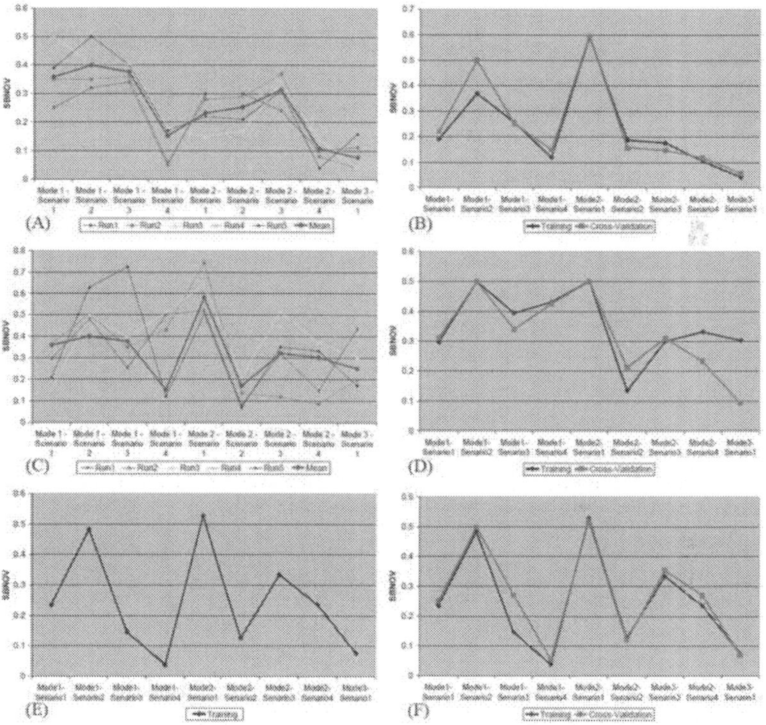

Figure 10. (A, C and E): SBNOV for the training data using GA, PSA and EK respectively. (B, D and F) SBNOV for the validation data using GA, PSA and EK, respectively.

In contrast, PSA produced relatively inconsistent results, which led to difficulties in observing the accuracy of each scenario. In addition, GA and EK have a similar pattern of accuracy across scenarios, with little variation in magnitude. For example, they gained similar levels of SBNOV accuracy in Mode 1—Scenario 1, Mode 1—Scenario 2, Mode 1—Scenario 4, Mode 2—Scenario 1, Mode 2—Scenario 2 and Mode 3—Scenario 1 but differ slightly in the remaining scenarios. This suggests that the GA was capable to some extent to understand the urban growth process and the underlying relationship between input factors in a way similar to human experts. It also suggests that the two algorithms have similar agreement about the efficiency of scenarios in terms of modelling urban growth. In contrast, the results of the PSA do not show results corresponding to those of the GA or EK. This might suggest that the complexity of the urban process is beyond the algorithm's capability as will be seen when we come to assess the accuracy of results.

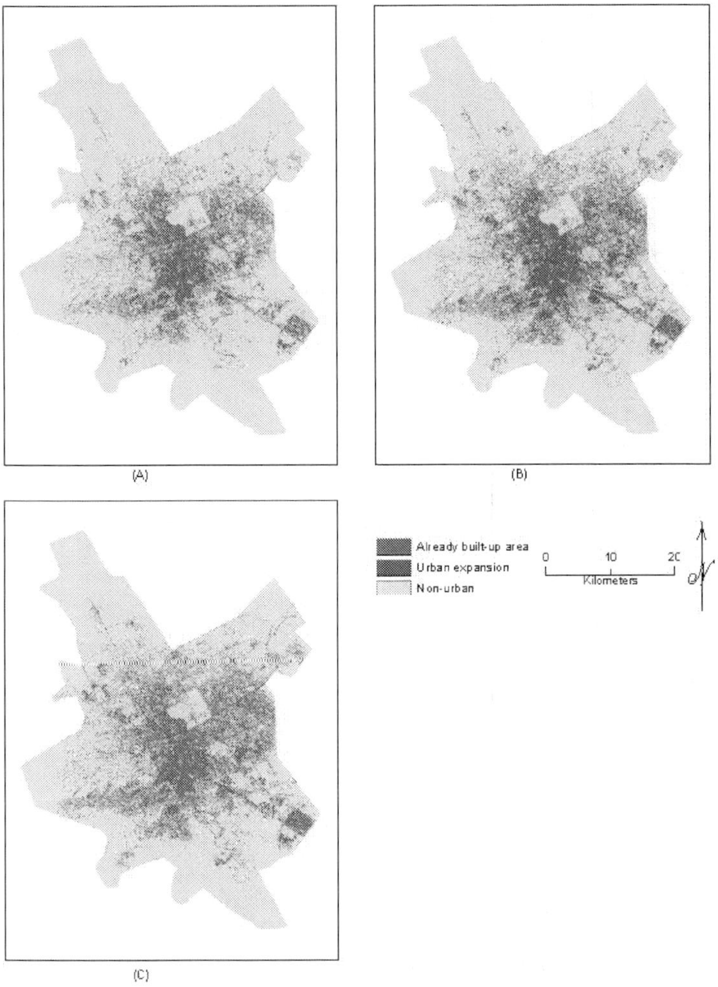

Figure 11. (A–C) Urban observed map (UOM) depicting urban expansion of Riyadh city during three periods: (A) 1987–1997, (B) 1997–2005 and (C) 1987–2005.

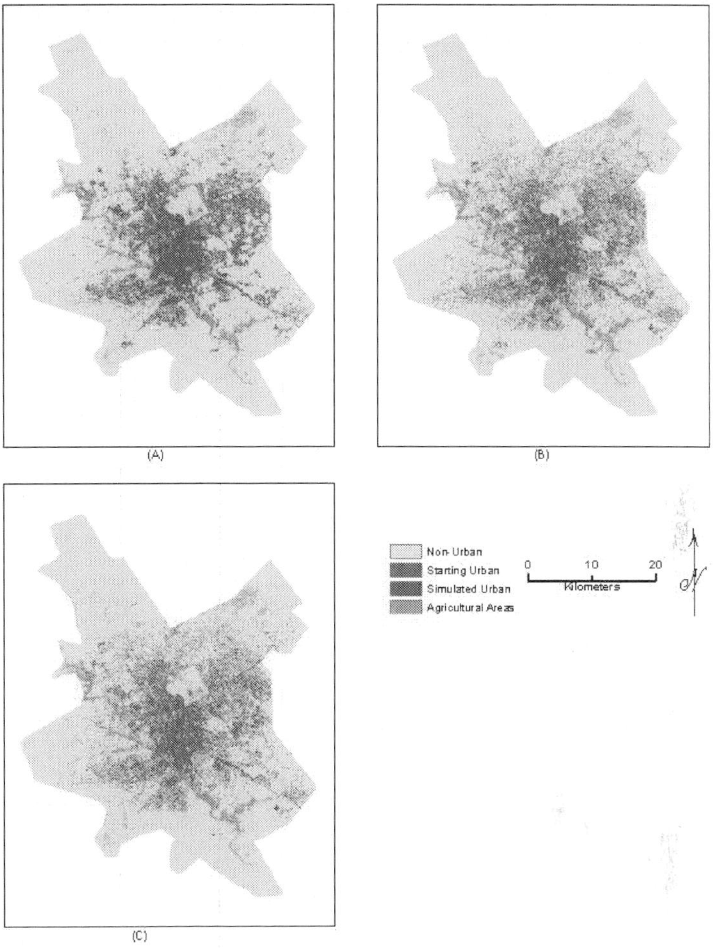

Figure 12. Performance of SIM-FCUGM for UGB I (1987–1997) period: (A) simulated image for scenario M1—S4; (B) simulated image for scenario M2—S4 and (C) simulated image for scenario M3—S1.

With respect to the accuracy of scenarios, it can be seen that Mode 3—Scenario 1 produced the higher levels of accuracy across all three algorithms, while Mode 2—Scenario 1 generated the worst solution. The high accuracy of Mode 3 might be attributed to the structure of this scenario, which includes three fuzzy variables in each fuzzy rule, that is, each fuzzy rule includes all of the three urban growth factors (TSF, UAAF and TCF). In addition, this high accuracy of Mode 3—Scenario 1 agrees with the results of uncertainty and sensitivity analysis, in which this scenario had the lowest uncertainty compared with others. The worst solution was produced by Mode 2—Scenario 1. This might be related to two factors: (i) the structure of the scenario and (ii) the type of driving forces employed in this scenario (which are TSF and TCF), that is, these two forces are not capable in this scenario of understanding the urban process of Riyadh. The low performance of this scenario is also revealed in the uncertainty and sensitivity analysis, indicating a weakness in structure of this scenario. **Figure 11** shows the urban observed map for 1987, 1997 and 2005, while **Figures 12–14** show the simulated urban growth during the three periods UGB1: 1987–1997, UGB2: 1997–2005 and UGB3: 1987–2005, respectively, that generated from THE best scenarios: M1—S4, M2—S4 and M3—S1.

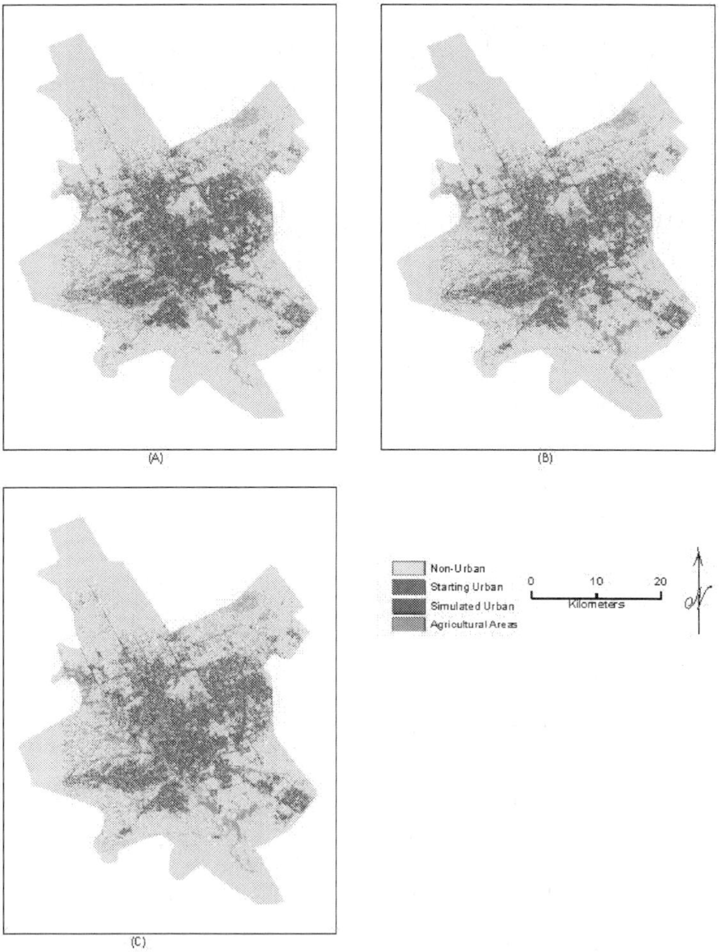

Figure 13. Performance of SIM-FCUGM for UGB I (1987–2005) period: (A) simulated image for scenario M1—S4; (B) simulated image for scenario M2—S4 and (C) simulated image for scenario M3—S1.

http://dx.doi.org/10.5772/64788

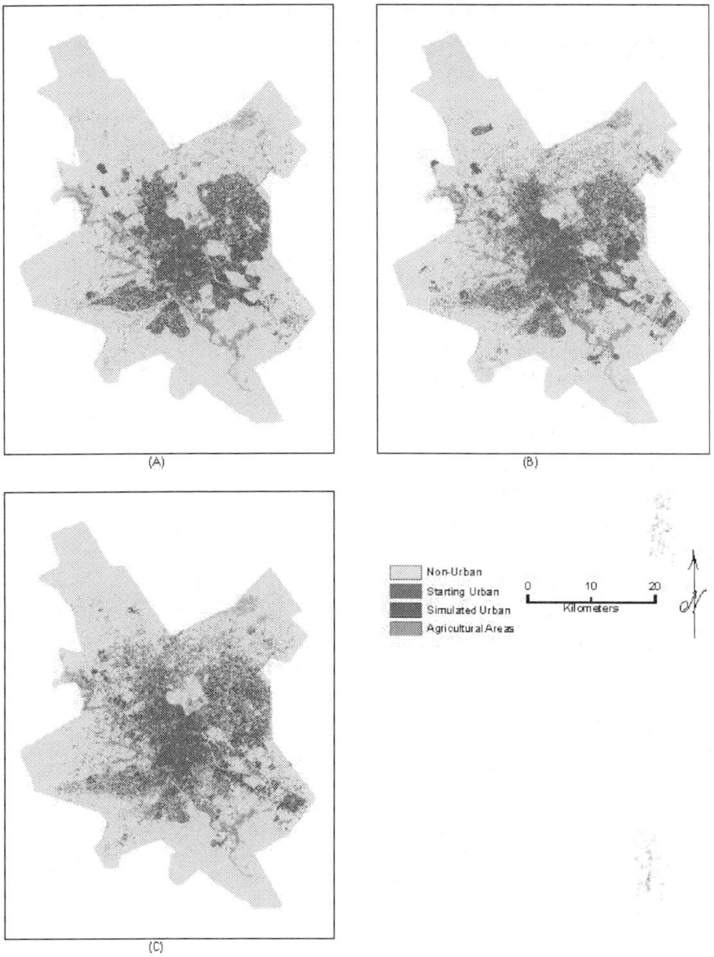

Figure 14. Performance of SIM-FCUGM for UGB I (1997–2005) period: (A) simulated image for scenario M1—S4; (B) simulated image for scenario M2—S4 and (C) simulated image for scenario M3—S1.

In relation to the validation of the calibration results, it can be observed that the GA and EK show validation results that are very close to one another and correspond closely to the training results, whereas the PSA presents lower matching results. For example, the GA and EK have identical training and validation results in all scenarios except Mode 2—Scenario 1 and Mode 2—Scenario 3 in GA and EK, respectively. In the PSA, only four scenarios match the training results including: Mode 1—Scenario 1, Scenario 2, Scenario 4 and Mode 2—Scenario 1, and the remaining five scenarios contradict one another. This implies that the GA and EK are better than the PSA, indicating that they have the capability to work well not only for the data that they trained with but with other data sets. Thus, in terms of generalization, it might be deduced that the CALIB-FCUGM by using the GA or EK can be used to calibrate different data sets from different times and locations.

7. Conclusion

In this chapter, theory underlying the CALIB-FCUGM has been applied to calibrate the FCUGM for Riyadh in Saudi Arabia. This chapter can broadly be divided into three main parts: uncertainty and global sensitivity analysis; calibration of the FCUGM; and results and discussion of calibrating the FCUGM. This chapter began by undertaking uncertainty and global sensitivity analysis on the scenarios in the FCUGM, which showed that the different structures of scenarios have different levels of uncertainty. It was found that Mode 3—Scenario 1, Mode 2—Scenario 4 and Mode 1—Scenario 4 generated the best performance, with the lowest uncertainty values, where 90% of the occurrences (iterations) of the Monte Carlo simulation for those scenarios gained the lowest error in terms of the objective function of the CALIB-FCUGM. After that, the technical stages of the calibration of the FCUGM were examined. These included the feasible solution, objective function, experimental design and calibration data set. This was followed by outlining the detailed processes of the optimization algorithms (GA, PSA and EK) within the CALIB-FCUGM. Next, empirical experiments were conducted to investigate the best control parameters of the GA and PSA for the FCUGM problem. It was found that the best GA and PSA parameters for the FCUGM problem had some similarity but differed with respect to problem in geography and non-geography. Finally, the FCUGM was calibrated under nine scenarios over three periods using three optimization algorithms. It was revealed that scenarios Mode 3—Scenario 1, Mode 2—Scenario 4 and Mode 1—Scenario 4 produced the best performance among the nine scenarios; this result is similar to that found in the uncertainty and global sensitivity analysis. The first reason for this is that the driving forces (TSF, UAAF or TCF) were embedded in those scenarios. This indicated that the spatial patterns of urban growth for Riyadh can be better understood by the three forces all together. The second reason can be attributed to the structure of the fuzzy transition rules, for example, Mode 3—Scenario 1, embedded all the three driving forces in each fuzzy rule and produced the most accurate results compared with others scenarios where their rule structure embedded only one or two driving forces.

It was found that the GA followed by EK produced better and more accurate and consistent results compared with PSA. This suggests that the GA was able to some extent to understand the urban growth process and the underlying relationship between input factors in a way similar to human experts. It also suggests that the two algorithms (GA and EK) have similar agreement about the efficiency of scenarios in terms of modelling urban growth. In contrast, the results of the PSA do not show results corresponding to those of the GA or EK. This suggests that the complexity of the urban process is beyond the algorithm's capability or could be due to being trapped in local optima. Investigation into the CALIB-FCUGM results over different urban growth periods indicated that, where the spatial pattern is more compact, the calibration results are more accurate. The calibration results over the period UGB I + II followed by UGB I produced better results compared with the one over UGB II. This can be understood due to the characteristics of the spatial pattern of urban growth for each period. UGB I+II followed by UGB I experienced edge expansion (relatively compact pattern), while UGB II faced in-filling development (dispersed compact pattern).

To sum up, CALIB-FCUGM was to a large extent able to calibrate the FCUGM over different growth periods under different scenarios using different algorithms. Although some algorithms and scenarios showed average performance, others revealed high capability for calibrating the model well. With this satisfactory calibration of the FCUGM for the urban growth of Riyadh by using CALIB-FCUGM, these calibrated parameters will be passed into the SIM-FCUGM to simulate the spatial patterns of urban growth of Riyadh.

Acknowledgements

The authors acknowledge with gratitude King Abdulaziz City for Science and Technology for the accomplishment of this work.

Author details

Khalid Al-Ahmadi*, Mohammed Alahmadi and Sabah Alahmadi

*Address all correspondence to: alahmadi@kacst.edu.sa

Space and Aeronautics Research Institute, King Abdulaziz City for Science and Technology, Riyadh, Saudi Arabia

References

[1] Allen P. Cities and Regions as Self-Organizing Systems: Models of Complexity. Amsterdam: Gordon Breach Science; 1997.

[2] Batty M, Longley PA. Fractal Cities: A Geometry of Form and Function. San Diego: Academic Press; 1994.

[3] Wilson A. Complex Spatial Systems: The Modeling Foundations of Urban and Regional Systems. Harlow: Pearson Education; 2000.

[4] Portugali J. Self-organization and the City. Berlin: Springer-Verlag; 2000.

[5] Batty M. The emergence of cities: complexity and urban dynamics [Internet]; 2003. Available from: http://discovery.ucl.ac.uk/231/1/paper64.pdf [accessed: 2006-04-12].

[6] Wolfram S. Cellular automata as models of complexity. Nature. 1984;311(5985):419–24.

[7] Toffoli T, Margolus N. Cellular Automata Machines: A New Environment for Modeling. Cambridge: The MIT Press; 1987.

[8] Batty M, Xie Y. From cells to cities. Environment and Planning B: Planning and Design. 1994;21(7):31–48.

[9] Wu F. A linguistic cellular automata simulation approach for sustainable land development in a fast growing region. Computers, Environment and Urban Systems. 1996;20(6):367–87.

[10] Wu F. An experiment on the generic polycentricity of urban growth in a cellular automatic city. Environment and Planning B: Planning and Design. 1998;25(5):731–52.

[11] Wu F. Simulating urban encroachment on rural land with fuzzy-logic-controlled cellular automata in a geographical information system. Journal of Environmental Management. 1998;53(4):293–308.

[12] Wagner DF. Cellular automata and geographic information systems. Environment and Planning B: Planning and Design. 1997;24(2):219–34.

[13] Batty M, Xie Y, Sun Z. Dynamics of urban sprawl [Internet]; 1999. Available from: http://discovery.ucl.ac.uk/1360/1/paper15.pdf [accessed: 2005-07-15].

[14] Batty M. Cities and Complexity: Understanding Cities through Cellular Automata, Agent-based Models and Fractals. Cambridge: The MIT Press; 2005.

[15] Guan D, Li H, Inohae T, Su W, Nagaie T, Hokao K. Modeling urban land use change by the integration of cellular automaton and Markov model. Ecological Modelling. 2011;222(20):3761–72.

[16] Al-shalabi M, Billa L, Pradhan B, Mansor S, Al-Sharif AA. Modelling urban growth evolution and land-use changes using GIS based cellular automata and SLEUTH models: the case of Sana'a metropolitan city, Yemen. Environmental Earth Sciences. 2013;70(1):425–37.

[17] Feng Y, Liu Y, Batty M. Modeling urban growth with GIS based cellular automata and least squares SVM rules: a case study in Qingpu–Songjiang area of Shanghai, China. Stochastic Environmental Research and Risk Assessment. 2016; 30 (5):1387–1400.

[18] Wu F. Calibration of stochastic cellular automata: the application to rural-urban land conversions. International Journal of Geographical Information Science. 2002;16(8): 795–818.

[19] Al-Ahmadi K, Heppenstall A, Hogg J, See L. A fuzzy cellular automata urban growth model (FCAUGM) for the city of Riyadh, Saudi Arabia. Part 1: model structure and validation. Applied Spatial Analysis and Policy. 2009;2(1):65–83.

[20] Al-Ahmadi K, Heppenstall A, Hogg J, See L. A fuzzy cellular automata urban growth model (FCAUGM) for the city of Riyadh, Saudi Arabia. Part 2: scenario testing. Applied Spatial Analysis and Policy. 2009;2(2):85–105.

[21] Al-Ahmadi K, See L, Heppenstall A, Hogg J. Calibration of a fuzzy cellular automata model of urban dynamics in Saudi Arabia. Ecological Complexity. 2009;6(2):80–101.

[22] Al-Ahmadi K, See L, Heppenstall A. Validating spatial patterns of urban growth using cellular automata. In: Salcido A, editor. Emerging Applications of Cellular Automata. Rijeka: InTech; 2012. p. 23–52.

[23] Yeh A, Li X. Error propagation and model uncertainties of cellular automata in urban simulation with GIS. In: The 7th International Conference on GeoComputation, Southampton, UK; 2003.

[24] Kocabas V, Dragicevic S. Assessing cellular automata model behaviour using a sensitivity analysis approach. Computers, Environment and Urban Systems. 2006;30(6):921–53.

[25] Saltelli A, Tarantola S, Campolongo F, Ratto M. Sensitivity Analysis in Practice. A Guide to Assessing Scientific Models. Chichester: John Wiley & Sons; 2004.

[26] Teknomo K. Monte Carlo Simulation Tutorials [Internet]. 2005. Available from: http://people.revoledu.com/kardi/tutorial/Simulation/index.html [accessed: 2006-08-10].

[27] Yeh A, Li X. Errors and uncertainties in urban cellular automata. Computers, Environment and Urban Systems. 2006;30(1):10–28.

[28] Wong D, Lee J. Statistical Analysis of Geographic Information: With ArcView and ArcGIS. New Jersey: John Wiley & Sons; 2005.

[29] Wu F, Webster C. Simulation of land development through the integration of cellular automata and multicriteria evaluation. Environment and Planning B: Planning and Design. 1998;25(1):103–26.

[30] Li X, Yeh AG-O. Calibration of cellular automata by using neural networks for the simulation of complex urban systems. Environment and Planning A. 2001;33(8):1445–62.

[31] Li X, Yeh A. Neural-network-based cellular automata for simulating multiple land use changes using GIS. International Journal of Geographical Information Science. 2002;16(4):323–43.

[32] Kim B. Modeling New City Growth Using Cellular Automata [Unpublished PhD Thesis]. Leeds: University of Leeds; 2005.

[33] Heppenstall AJ. Applying of Hybrid Intelligent Agents to Modeling a Dynamic, Locally Interaction Retail Market [Unpublished PhD Thesis]. Leeds: University of Leeds; 2004.

[34] Pontius R, Schneider L. Land-cover change model validation by an ROC method for the Ipswich watershed, Massachusetts, USA. Agriculture, Ecosystems & Environment. 2001;85(1):239–48.

[35] Kirkby M, Naden P, Burt T, Butcher D. Computer Simulation in Physical Geography. Chichester: John Wiley & Sons; 1993.

[36] Rogreson PA. Statistical Methods for Geography. London: SAGE Publication Ltd.; 2006.

[37] Whitley D. A genetic algorithm tutorial. Statistics and Computing. 1994;4(2):65–85.

[38] Mitchell M. An Introduction to Genetic Algorithms. Cambridge: The MIT Press; 1998.

[39] Reeves CR, Rowe JE. Genetic Algorithms – Principles and Perspectives: A Guide to GA Theory. Norwell: Kluwer Academic Publishers; 2003.

[40] Van Laarhoven PM, Aarts EH. Simulated Annealing: Theory and Applications. Netherlands: Springer; 1987.

[41] Cheng J, Masser I. Understanding spatial and temporal processes of urban growth: cellular automata modelling. Environment and Planning B. 2004;31(2):167–94.

[42] Ward DP, Murray AT, Phinn SR. A stochastically constrained cellular model of urban growth. Computers, Environment and Urban Systems. 2000;24(6):539–58.

Chapter 5

Structural Diversity of Plant Populations: Insight from Spatial Analyses

Janusz Szmyt

Additional information is available at the end of the chapter

http://dx.doi.org/10.5772/65320

Abstract

Spatial analysis has been one of the most rapidly growing fields in ecology. It is related directly to a growing awareness among researchers that a spatial structure of biosystems, e.g., forests, is important in ecological thinking. The availability of the specific software supports the use of spatial analyses in different fields of the science and forestry science is only one example for this. Many data collected in the forests have the spatial and temporal dimensions and it allows us to use spatial statistics to quantitative description of the spatial structure of forest, which became an important element of modern continuous cover forestry. In this chapter, key elements: data types, null models, and summary statistics, which can be applied in spatial analyses, are briefly described. Real data sets collected from different forests were given to provide examples of spatial analyses. The key elements of spatial analysis in ecology are data type, the appropriate choice of summary statistics and null models. Selecting few of them in a single analysis makes the statements more reliable and realistic in the changing world.

Keywords: spatial explicit indices, spatial functions, point pattern statistics, forest structure diversity, forest dynamics

1. Introduction

Ecologists have been interested in spatial and temporal dimensions of ecological processes in plant populations for a long time. While data collected in most ecological studies have spatial and temporal aspects, the importance of spatio-temporal analysis has been discovered recently. As stated in Reference [1], until the 1980s, most ecological studies avoided the explicit consideration of space and most of the field experiments were designed to remove spatial signals. Techniques such as randomization and block designs were especially common in use.

During the 1980s, there was a fundamental shift in ecology toward spatial explicit consideration of relationships between organisms. Among factors favorable to use spatial analysis in ecological studies one can distinguish the needs to include spatial structure of natural systems in ecological theories, changes in landscapes altering ecosystems, and the needs to evaluate their spatial heterogeneity and—which was most influential—to develop a modern technology which increase possibilities in analyzing a large spatio-temporal data sets together with the development of specific statistical methods (e.g., point process statistics), technology (e.g., LIDAR), and software dedicated to spatial analysis [2]. The third factor allowed to analyze, model and visualize a complex spatial relationships between organisms even in rather complex biosystems, like tropical forests. Thus, at present, spatial analysis has been one of the most rapidly growing fields in ecology and it is now related directly to a growing awareness among researchers that spatial structure of populations (e.g., forest trees) is important in ecological thinking.

An important concept related to biological structures includes self-organization, structure relations, and pattern recognition [3]. Self-organization involves a variety of interactions between individuals (e.g., competition, facilitation), which can modify their growing spaces and spatial niches. Ecological processes leave signs in the form of spatial patterns but the spatial structure of the system can determine its properties, as well. In a forest, for example, population structure affects the biomass production, biodiversity, and habitat functions. Pattern recognition plays thus an important role in forest ecology and usually helps to identify and link spatial patterns with corresponding properties of population [1, 4–7].

The questions tried to be answered on the basis of spatial analysis often revolve around identifying the potential causes, e.g., ecological processes and mechanisms, staying behind the observed arrangement of individuals in the population [1, 8]. Historically, spatial analysis based on point pattern statistics provided only the assessment whether the empirical pattern of the studied population emerged by chance, which meant that individuals' occurrence did not depend on the presence of others, and the probability of the occurrence was the same across the whole study area. This expectation is called complete spatial randomness (CSR). Two alternatives to CSR are individuals that are distributed according to the specific mechanisms promoting either their overdispersion (aggregations, clumping) or underdispersion (regularity) [1, 8]. Nowadays, modern spatial statistics, e.g., point pattern analysis, allows us to find out more detailed information on spatial relationships between individuals in the investigated population. Some complex null models, such as Cox and Gibbs processes, can be helpful for that. In general, cluster models of Thomas, Neyman-Scott, and Matern, being representatives of Cox processes, provide detailed information on the average cluster size and the number of clusters per unit area. On the other hand, the Gibbs class of point process models (e.g., Strauss and Markov processes) can characterize inhibition mechanisms between individuals [8]. Point process models, mentioned above, are important tools employed in spatial analyses. Their importance results from their usefulness in determination weather there is any significant spatial structure in empirical data, they can summarize the properties of the spatial structure and test ecological hypotheses concerning the mechanisms that may generate the observed spatial structure in a data set [1].

Fundamental ecological questions arising in forestry concern the forest structure and its influence on forest dynamics, forest productivity, and biodiversity [9–12]. This refers to the way in which the attributes of trees (species, sizes) are distributed in the forest.

It affects most ecological processes running in the forest ecosystem, among which forest regeneration, tree growth, surviving and mortality, seed dispersal, competition, or facilitation between individuals are especially important (**Figure 1**). Moreover, most of biological processes generate themselves the specific structures. Thus, the structure-processes relations are not independent. Forest dynamics depends thus to a large degree on the forest structure.

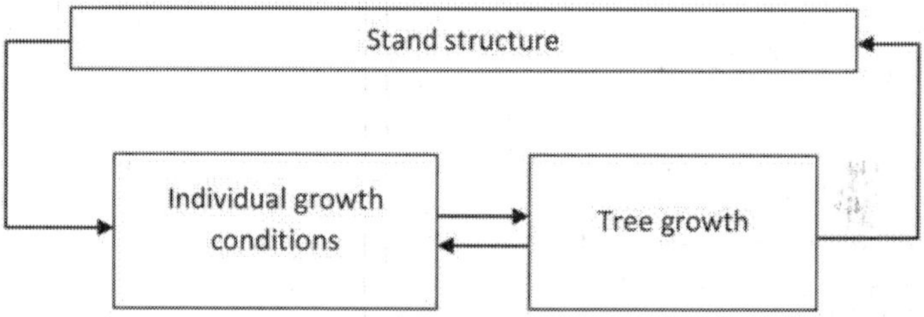

Figure 1. Feedback loop determining forest stand dynamics [9, modified].

This chapter is divided into the following subchapters:

1. Data types—what should be known before running the spatial analysis.
2. Patterns and processes—the mutual dependence causes some inferential problems.
3. Spatial indices—an easy way to describe population structure.
4. Functional spatial statistics—the most informative way to discover complex structures.
5. Conclusion

2. Data types—what should be known before running the spatial analysis

Generally, the aim of spatial analysis is to describe the structure of the pattern created by objects distributed in space. Each object is usually treated as a point, regardless their real shapes and point pattern statistics seem to be valuable tools in such analysis.

As mentioned above, most data collected in ecological studies can be characterized by spatial dimensions. However, data can be of different types and selection of the appropriate statistical method (the so-called summary statistics) depends on two things: the data we want to analyze and ecological questions we want to answer [8, 13]. Individuals being the subjects of spatial analysis are usually characterized by their location (x, y coordinates) and additionally by their different attributes, quantitative, or qualitative ones (e.g., size, species, sex, quality, health status, and age). It is also possible to use as a tree attribute any constructed mark [14].

Individuals described only by coordinates can be analyzed as the so-called unmarked point pattern, while data described by any mark are suitable to analyze as the marked point pattern [8, 15]. The appropriate summary statistics (indices and functions) that quantify the statistical properties depend on the form of the data type one collected in the field. Another important issue associated with the point pattern analysis is the heterogeneity of environment conditions. In ecology, heterogeneity plays an important role and its quantification seems to be a key task in spatial analysis. To do that, the information on environmental covariates (soil quality, slope, aspect, etc.) should be incorporated in analysis [16].

In the unmarked point pattern analysis, one would like to characterize the spatial relationships between objects, e.g., trees in the forest. Moreover, the unmarked pattern may include one or more types of individuals. The analysis of such point patterns concerns the following basic categories: univariate, bivariate, and multivariate point patterns [1, 15]. Univariate point pattern analysis is focused only on one type of points, e.g., particular tree species. The questions to be answered are about the understanding of the mechanisms (processes) responsible for the distribution of the individuals within the study area. The fundamental null model for the univariate analyses is the complete spatial randomness and it is called the (homogeneous) Poisson model. According to CSR, points are distributed with equal probability within the region of interest and each point is distributed independently of the others. The alternatives to CSR are, either aggregated or hyperdispersed arrangement of points. In the case of the bivariate point pattern, two types of points are the subjects of analysis. It is important to keep in mind that these two types of points must be created by two different processes [8]. Such points have the so-called a priori properties [16]. Good examples of bivariate point patterns in forest studies are analyses of spatial correlation between two different tree species or live stages (adults and juveniles). In the case of the bivariate pattern, the null model is spatial independence of two patterns and the alternatives are spatial attraction (positive association) and spatial repulsion/segregation (negative association). The main question is focused on the role of interactions between two types of points. Bivariate analysis can support the theory of species coexistence in multispecies forests [17–22]. In multivariate point pattern analysis, several data types (e.g., tree species) are involved and each of them is created by different processes. The relevant ecological questions for such data types involve detecting and understanding spatial structures in diversity, namely whether tree species tend to form intraspecific and interspecific structures or different tree species tend to be well mixed over the study region. According to the hypothesis of spatial segregation in promoting the species coexistence, for example, intraspecific clusters for a certain species are responsible for the interspecific segregation [1, 16, 23].

In spatial analysis mentioned above, points of similar or different type were characterized only by their location. If we describe each point additionally by any mark (e.g., tree diameter, tree height, and health status), we obtained qualitatively or quantitatively marked patterns. Thus, summary statistics from the so-called marked point pattern statistics should be used. Qualitative marks are usually created by the a posteriori marking process over the given point pattern. This situation is quite different from the case of the bivariate pattern, created by a priori process. In the case of a qualitatively marked point pattern, one is interested in the characteristics of the process distributing the marks over the pattern. The relevant null model for qualitative marks is a random labeling (or independent marking) model, in which marks are shuffled in a random way over the joined pattern [1, 15]. In the case of quantitative marks, the relative ecological questions are about the spatial correlation of marks created a posteriori, too [7, 24, 25]. Such analysis can reveal, for example, the importance of competition (or cooperation) between trees in the dependence on the distance they are apart from each other.

Figure 2 presents major characteristic of the forest structure and its important variables.

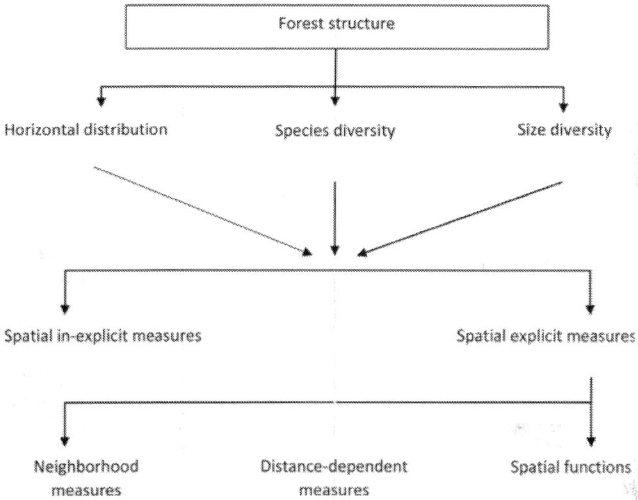

Figure 2. Major characteristics of the forest structure and its measures.

3. Patterns and processes: complex mutual dependence

As mentioned above, the natural processes and mechanisms leave some traces in the spatial pattern of individuals occupying a certain area [6]. These traces encompass different aspects of population structure: species composition and species mixing, spatial arrangement of individuals and spatial variation of their size [26, 27]. To understand the functional processes it is needed to identify the structure and spatial scales at which processes operate. Spatial patterns in plant populations, e.g., forests, determine their integrity, functionality as well as stability to the large extent [1, 5, 9, 10, 16, 26, 28].

In ecological studies, there are numerous examples of the attempts of inference the underling processes from the observed patterns (structures). Spatial patterns of any population can be treated as an "ecological archive" in which the past ecological processes are conserved [16]. Decoding the signals from spatial patterns is still challenging due to the complex relationships between the pattern and the structure of plant population. Some potential problems arise from the fact that different processes can generate the same spatial pattern or they may interact. The processes may also be the result of the specific spatial patterns (spatial structures). Moreover, nonrandom processes can also generate random pattern [1, 6, 9, 27–31]. The inverse situations—that means a nonrandom process can create structured patterns—can be true either. Different processes do not have to interact simultaneously and a single process can generate exactly a single pattern [32].

The appropriate use of null models in spatial analyses, as well as complete description of the properties of the observed spatial pattern, allows us to minimize the problems stated above. One possibility to solve them is the use of several summary statistics simultaneously. The more structured population, the more number of summary statistics should be used in description of the pattern [33]. However, the use of a single or two summary statistics are the most common in the literature [16]. Historically, only a single null model, namely CSR, was used to state if the population is randomly distributed or not. Now, there are much more null models available for better analysis [8, 34–37].

In forests, spatial patterns revealed by trees are usually the result of three main biological processes: tree growth, mutual interactions, and mortality [14]. All these factors influence the forest dynamics and also its structure at the subsequent forest development stages. Tree growth can be impeded or "accelerated" due to different ecological processes and the neighborhood effect is among others [32]. Competition processes are difficult to measure directly; however, its effect on the tree growth and survival can be studied by spatial pattern analysis. Distance-dependent mortality of trees has been quite frequently referred to as a consequence of density-dependent competition, and this process frequently leads in crowded population to a more regular distribution of surviving trees [4, 38–40]. The relationships between small and large trees may be more complex. Small trees may tend toward aggregation around large trees because of better moisture conditions around larger trees or they tend to be segregated from large individuals because of poor light regimes for their growth and development [41]. In multispecies forests, interspecific competition may be reflected by spatial segregation of different tree species, and it is extremely important for weaker competitors allowing them to survive [42]. Thus, heterospecific segregation promotes species coexistence in mixed forests [1].

4. Spatial indices: an easy way to describe population structure

Spatial explicit indices can be divided into three main groups: quadrat counts, distance-based, and angle-based indices. Great advantage of the use of spatial indices is related to the fact that they are easy to calculate and results can be interpreted easily. However, the use of indices usually does not allow to draw conclusions on the spatial pattern of individuals at different spatial scales, but results can be interpreted only at a single scale, e.g., nearest neighborhood [3].

4.1. Quadrat counts

A quadrat counts method is based on counting points in subareas (quadrats) located in the particular region of interest [8, 43, 44]. This method is the oldest and the simplest measure of the pattern and intensity of population. The simplicity results from the fact that only the number of objects (trees) in a quadrat is obtained and there is no need to know the exact position of them. However, it limits the statistical analysis. The disadvantage of quadrat counts method is that the dispersion of the objects may depend on the scale of the study and the size of the sample unit [37, 43].

4.1.1. Variance-mean index (VM)

The most common index that can be applied to quadrat counts is the index of dispersion, also called the variance-mean ratio, and being based on the Poisson distribution. For the random distribution of points (followed the Poisson distribution), the index VM = 1. If points are aggregated then VM > 1 and if they are evenly scattered, thus regularly distributed, the index VM < 1 [43, 45–47]. In the first case, the variability in the process is stronger than in the Poisson process, and in the second case—the variability is smaller. For statistical inferences about the significance of the deviation from 1 (randomness), χ^2 test for $n - 1$ degrees of freedom can be used (n is the number of quadrats).

4.1.2. Morisita index (I_M)

Another easy-to-calculate index related to the quadrat counts method is Morisita's index, I_M, calculated from the number of objects on the squares, the number of squares and the total numbers of individuals [9, 43]. The standardized index takes the values from $I_M \epsilon \{-1, 1\}$ using either of two values calculated from χ^2 test with $n - 1$ degrees of freedom. If $I_M < 0$ then points within the population are distributed regularly, while $I_M > 0$ indicates the aggregated spatial structure [43]. Random distribution of individuals is for $I_M = 0$. The standardized index is assumed to be a very good measure of the spatial pattern because it is not affected by the population density and sample size. This index was applied in References [48–51].

Example 1

To illustrate the application of the Morisita index, data sets from an old-growth oak-dominated (*Quercus robur* L.) forest, located in western Poland will be used. **Figure 3** presents the stem map of the forest. Only hornbeam (*Carpinus betulus* L.) was taken into consideration for IM calculations.

The dependence of the spatial point pattern on the spatial scale on the basis of the Morisita index is presented in **Figure 4**. The pattern was divided into 2·2 quadrats, then 3·3, 4·4, etc., I_M index was calculated for each quadrat.

Results indicated that trees belonging to this species were distributed in clumps ($I_M > 1$), especially at small spatial scale. The larger spatial scale, the lower clumping intensity was observed.

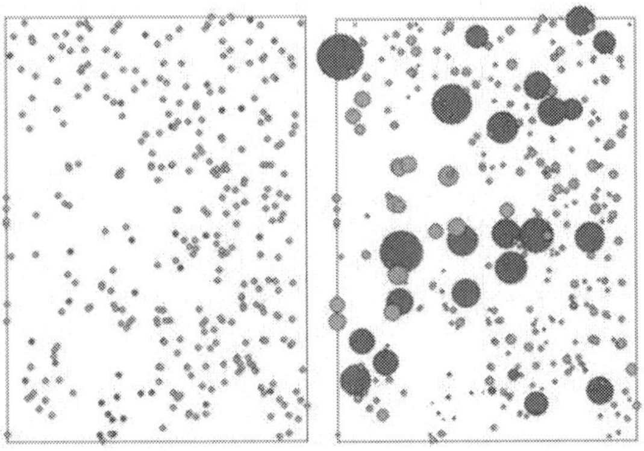

Figure 3. Stem map generated for oak-dominated (*Quercus robur* L.) old-growth forests (Example 1), the plot size: 50 m × 70 m. In the left panel: all live trees; red circles: pedunculated oak; green circles: hornbeam (*Carpinus betulus* L.). In the right panel: the size of circles corresponds to the diameter of each tree.

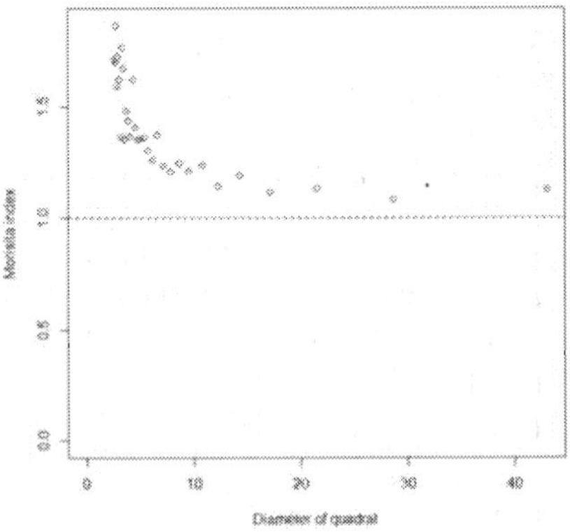

Figure 4. Values of Morisita index calculated for hornbeams in an old-growth oak-dominated forest and its dependence on the spatial scale. The point pattern is divided into quadrats of different size, and the Morisita index is computed each time. This plot discerns different scales of dependence in the point pattern data.

4.2. Point pattern statistics

Spatial point pattern analysis is based on the data sets consisted of objects with known locations. Modern ecological analyses are mainly based on point pattern (process) statistics and objects being the subjects of analysis are represented by points and marks describing them. In this subchapter, the most common and powerful methods are briefly described and they are supported by examples based on the real data sets from forest ecosystems. For the readers convenience, mathematical concepts are omitted in this chapter but they can be found in many textbooks on spatial statistics, e.g., in Refs. [1, 8, 37, 44, 52, 53].

4.2.1. Spatial arrangement

4.2.1.1. Distance-based indices

Spatial structure of a forest is largely determined by the relationships between close neighbors, thus, the neighborhood scale seems to be very important. A group of methods called the nearest neighbor statistics are based on the relative positions of individuals in the population [27]. Different indices from this group can provide the information on the different aspects of spatial structure: spatial arrangement of trees, spatial differentiation of their sizes, spatial mingling of tree species, etc. Some of them require an exact position of each tree in the population and the others require the position of only a sample trees. Distance within this group can be measured between the sample point to the nearest tree and from tree to its nearest neighbor [54].

4.2.1.1.1. Clumping index of Clark-Evans with Donnelly's modification (CE)

This index was introduced by Clark and Evans in 1954 and then it was modified using an edge correction formulae [55]. This index has been historically the most commonly used in spatial pattern analysis due to its simplicity and easy interpretation. The index is based on the distances between the nearest neighbors, measured for each tree within the population under investigation. It is a measure of the extent to which the population being analyze deviates from the random one. For randomly dispersed population CE = 1. If individuals are distributed in clumps then CE < 1, if they are dispersed regularly then CE > 1 [56] and for two alternative pattern type it is CE > 1 (regularity) and CE < 1 (aggregated). The maximum value of CE index is CE~2.15 for a hexagonal distribution of individuals [55–58]. The significance of the departures from 1 can be obtained by using a standard, normally distributed test value [59]. This author argued that the special attention with the application of the CE index should be drawn in populations where clustering is likely to be present. Then, other indices are assumed to provide more reliable results. Another weakness of the CE index is that it assumes that the process generating tree location is homogeneous and in the case of spatial variations of point density this index will show the virtual aggregation [37].

4.2.1.1.2. Hopkins-Skellam index of dispersion (HS)

This index, unlike CE, takes the nearest neighbor distances between the randomly sampled points and the random object of the pattern (e.g., tree). The pattern is random when points are independently distributed from each other and the distance from the data point to its nearest neighbor should have the same probability distribution as the distance from a fixed spatial location to the nearest point of the pattern [43, 37]. This index, similarly to the CE index, is dimensionless. For random population HS = 1, for aggregated structure HS < 1 and for regularly spaced individuals, HS > 1. The HS test compares the value of the index to the *F*-distribution. Hopkins-Skelam index is less sensitive than CE due to edge effect bias and spatial inhomogeneity [37].

4.2.1.2. Angle-based indices

Both indices described above require the measurement of the distances that is rather time consuming and laborious. For this reasons, two indices based on angles between nearest neighbors, namely, contagion index and mean directional index, have been introduced by Corral-Rivas et al. [60] and Aguirre et al. [61], respectively. Their basic idea is to characterize the spatial pattern of trees at the neighborhood scale by the directions under which the n neighbors of the so-called reference point were visible. Each point of the pattern takes a role of reference point.

4.2.1.2.1. Uniform angle index (also known as contagion index) (UAI)

This index is based on the classification of the angles α_{ij} (i refers to the reference tree and j to its neighbors) between two neighbors. It compares these angles with an appropriate reference angle, α_0, which is selected so that it yields 360°/n [10]. The contagion is defined as the proportion of angles α_{ij} between the four neighbors, which are smaller than α_0, and the index takes the values between 0 (regularity) and 1 (clumping). In the case of four neighbors, UAI can take five values: 0.0, 0.25, 0.5, 0.75, and 1.0. Mean values for a stand are an arithmetic mean of all UAI values calculated for each trees. Mean values of UAI > 0.6 indicate clumped distribution and UAI < 0.5—regularity [9, 10, 62]. More informative than the stand average value is the distribution of UAI that provides detailed information how many trees are arranged in clumps and how many trees are distributed randomly or regularly. As stated above, this index is a suitable tool when the number of points exceeds 100 individuals [61].

4.2.1.2.2. Mean directional index (MDI)

This index is more conventional that the previous one and more accurate angle measurements are necessary, but still no distances should be measured. Usually, values obtained by MDI index correspond well with values obtained by the UAI index. If trees are distributed in regular manner MDI = 0 and if they are distributed in clumps—MDI takes larger values. The mean MDI index for the stand can be also calculated. The value of the MDI index for a random population is exact 1.7999 (~1.8). Thus, values MDI > 1.8 indicate aggregated structure and MDI < 1.8—regular distribution of individuals. This index is suitable in the case of the populations with the number of individuals exceeding 50 objects [61, 63].

Example 2

The application of spatial indices is supported by real data set collected from old-growth oak-dominated (*Quercus robur* L.) forest, located in western Poland. **Figure 3** presents the stem map of trees in the forest located in the nature reserve in Poland. This forest has been excluded from any human interventions since the last 50 years. The main tree species is pedunculated oak (overstory), and hornbeam (*Carpinus betulus* L.) in the understory. The age of oaks was approximately 160 years and hornbeam ca. 70–90 years. Each tree was described by its coordinates and marks: diameter at the breast height (*dbh* in cm) and the total tree height (*h* in m). **Table 1** presents the values for nearest-neighbors indices (CE, HS, UAI and MDI) for all trees and for each tree species, separately.

Spatial measure	All trees	Pedunculate oak	Hornbeam
CE	0.90*	1.10	0.90*
HS	0.65*	0.89	0.56*
UAI	0.51	0.50	0.51
MDI	1.89*	1.99*	1.88*

*Significant departures from CSR at the $\alpha = 0.05$.

Table 1. Average values of spatial indices calculated for all trees in old-growth oak-dominated forest.

Both distance-based indices, CE and HS, clearly indicated clustering of all living trees. In the case of angle-based indices, only MDI was consistent with results obtained by distance-based ones. The UAI showed random distribution of living trees. Oaks showed random distribution and it was confirmed by CE, HS, and UAI indices but not by MDI. The latter showed their clumped distribution. The spatial pattern of hornbeam was also clumped and most indices confirmed that, except UAI. On the basis of obtained results, one can state that the spatial pattern of trees in the forest density of hornbeam, easily regenerated from sprouts.

4.2.2. *Spatial variation in size: spatially explicit size differentiation indices*

Apart from the spatial arrangement of trees, tree size differentiation is assumed to be an important characteristic describing population diversity. Two commonly applied spatial indices seem to be interesting: size differentiation index and (relative) dominance index.

4.2.2.1. *Size differentiation index (T)*

This index describes the similarity or dissimilarity of size of individuals being the nearest neighbors. The neighborhood of the reference tree consists of three or four neighbors of a reference tree. The *T* index is a single value calculated for each tree within the population and an arithmetic mean gives the information on the average size differentiation of trees in the forest. In extremely high structured population the value of $T = 1$, whereas in population where individuals are quite similar it is close to $T = 0$. The arithmetic mean provides the general insight into structural diversity of the forest, at the stand level. However, more informative is the share of trees belonging to the particular differentiation classes: 0–0.30, very small differentiation; 0.30–0.50, moderate differentiation; 0.50–0.70, high differentiation; 0.70–1.00, very high differentiation [10]. To find out if the departures from the expected value of *T* under the random conditions are statistically significant, a permutation procedure can be applied.

4.2.2.2. Size dominance index (D)

This index aims at the description of the relative dominance of a given tree to its nearest neighbors. It can be defined as the proportion of the n neighbors of a reference tree, which are smaller in size than the reference tree [62, 64]. If four neighbors are taken into consideration, D index can take again five values corresponding to different biosocial categories according to Kraft's crown classification: 0.00, very suppressed (all neighbors are smaller than the reference tree); 0.25, moderately suppressed; 0.50, codominant; 0.75, dominant; 1.00, strongly dominant (none of neighbors are smaller than the reference tree).

Example 3

Figure 5 presents the location of trees in a managed old-growth beech-dominated (*Fagus sylvatica* L.) forest. The main tree species was European beech and silver fir (*Abies alba* L.) was admixture species. Both tree species occurred in the overstory. The average age of the forest was 145 years. Up to the year of measurements, the forest stand has been managed according to Polish standards for beech stands. Apart from the location of each live tree in the stand (x, y coordinates), diameter at the breast height (dbh, in cm) and the total tree height (h, in m) were measured and tree species were reported.

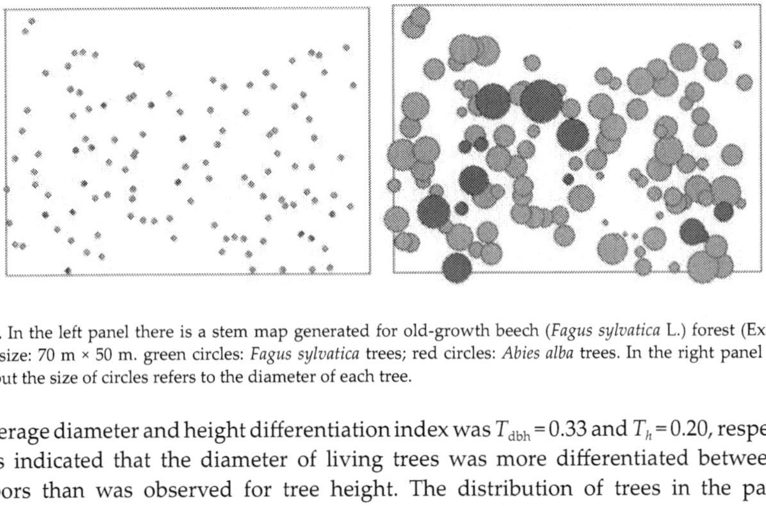

Figure 5. In the left panel there is a stem map generated for old-growth beech (*Fagus sylvatica* L.) forest (Example 3), the plot size: 70 m × 50 m. green circles: *Fagus sylvatica* trees; red circles: *Abies alba* trees. In the right panel the same pattern but the size of circles refers to the diameter of each tree.

The average diameter and height differentiation index was $T_{\text{dbh}} = 0.33$ and $T_h = 0.20$, respectively. Results indicated that the diameter of living trees was more differentiated between close neighbors than was observed for tree height. The distribution of trees in the particular differentiation classes showed that the neighbors of ca. 43% of trees were only slightly different in dbh, and 50% of trees was surrounded more differentiated individuals. In the case of tree height, the trend is similar but the differences between nearest neighbors are much less stressed (**Figure 6**).

The average spatial differentiation index calculated for diameter for beech and silver fir was $T_{\text{dbh}} = 0.32$ and $T_{\text{dbh}}=0.37$, respectively. In the case of tree height, these indices were $T_{\text{h}} = 0.19$ and $T_{\text{h}}= 0.26$ for beech and fir, respectively. **Figure 7** shows the distribution of trees in the particular size differentiation classes. Trees of both species showed more or less similar distribution in particular size differentiation classes in the case of both tree attributes.

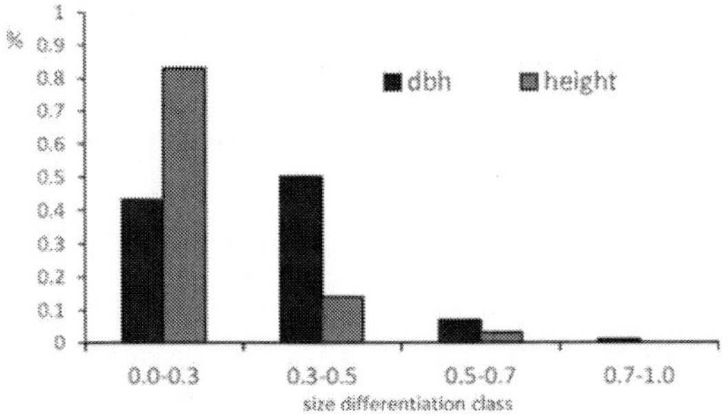

Figure 6. Distribution of live trees in size differentiation classes for diameter (dbh) and total tree height in an old-growth *Fagus sylvatica* forest (Example 3).

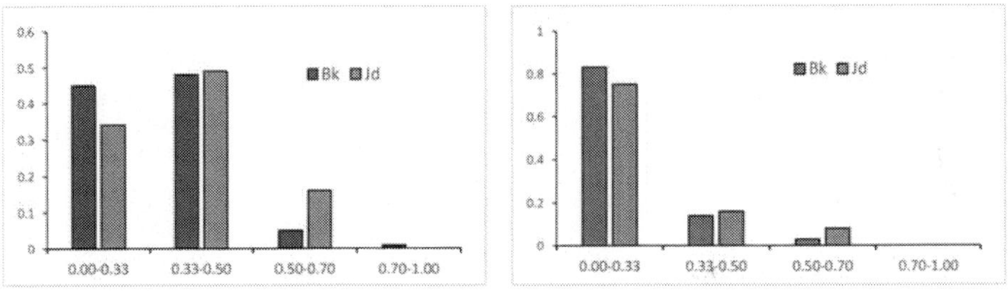

Figure 7. Distribution of trees of different species in size differentiation classes for dbh (left panel) and total tree height (right panel) in an old-growth *Fagus sylvatica* forest (Example 3).

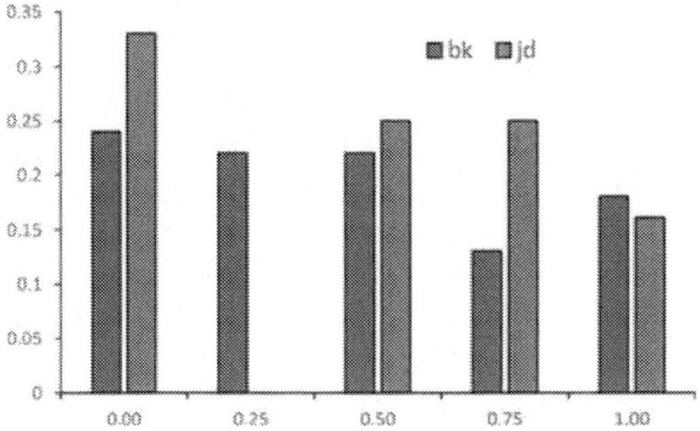

Figure 8. Dominance distribution of European beech (Bk) and silver fir (Jd) in an old-growth beech-dominated forest.

Trees showed small to moderate diameter differentiation in the neighborhood scale (ca. 90% of trees). At the same time, height differentiation of nearest neighbors was clearly lower and most trees showed small differentiation around (ca. 83% of trees) (**Figure 6**). In general, the diameter was more differentiated than the tree height for both tree species in the forest.

Dominance criterion is useful for describing the relative dominance of different tree species, for example European beech and silver fir from example data set presented here. The distribution of beech is left-skewed meaning that the majority of trees of this tree species are surrounded by at least three bigger neighbors. However, there are few dominant beech trees. Similar constellation was observed in the case of silver fir (**Figure 8**).

4.2.3. Spatial mixing of species

The third aspect of spatial structure is attributed to the relative mingling of different species in plant community. Two indices can be taken into consideration: species mingling index introduced by von Gadow and species segregation index introduced by Pielou [65].

4.2.3.1. Species mingling index (MI)

This index describes the spatial distribution of different tree species around the reference tree [10, 27, 64, 66]. It is determined for each individual (reference tree) within the population and it gives the proportion on the nearest neighbors (e.g., 4), which are not of the same species as reference tree is. The index takes values between 0 and 1 and if four neighbors are taken into account, five values of MI can be obtained: 0.0 (all neighbors are of the same species as reference tree), 0.25, 0.50, 0.75, and 1.0 (all neighbors are of different tree species as reference tree). Similarly to previously describe indices, the distribution of MI provides a more detailed insight into species composition of the forest. To find out whether departures from the random mixing are statistically significant, a permutation procedure can be applied.

4.2.3.2. Species segregation index (SSI)

This index describes the relative mixing of only two species regardless of their spatial pattern. If there are more than two species in the population, each pair of species should be analyzed separately. The SSI index is based on the comparison of the observed number of mixed species pairs and the expected number if the two species would be distributed independently of each other [9, 59, 67]. The SSI values can lie between −1 and 1. Two species are associated together (aggregated) if SSI < 0 and they are segregated if SSI > 0. They are randomly distributed from one another if SSI = 0 [59]. A χ^2 test may be applied to judge the significance of the departures from random mixing of both species.

Example 4

Let's go back to the oak-dominated old-growth forests introduced earlier (see Example 1). Two tree species are present in the stand. The average value for the mingling index (MI) is small, MI = 0.13, suggesting that tree species are distributed in a homogeneous patches. In the case of oak, MI = 0.40 indicating that they are distributed in heterogeneous clumps, while hornbeams are distributed in homogeneous patches (MI = 0.06).

http://dx.doi.org/10.5772/65320

As shown in **Figure 9**, the trees form mostly homogeneous patches. About 70% of trees are surrounded by the same tree species. It is caused mostly due to the hornbeam. About 80% of individuals of this tree species are surrounded by conspecifics. The surroundings of oaks are mostly heterogeneous and three of four neighbors of this tree species (70% of oaks) are of different species.

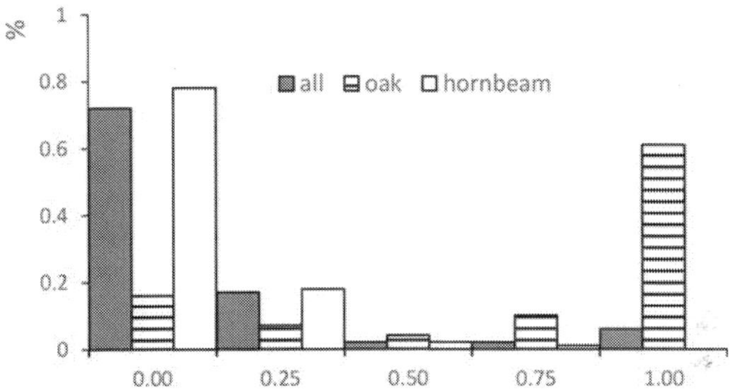

Figure 9. Distribution of species mingling index (MI) for all live tree, oaks and hornbeams in the oak-dominated old-growth forest.

Applying the Pielou's segregation index (SSI), we obtained only limited information on the probability to find individuals of one species in the neighborhood of the individuals of the other species. In the example, the SSI index showed random mixing of oak and hornbeam (SSI = 0.25, p-value = 0.25).

5. Functional spatial statistics: the most informative way to discover complex structures

A great advantage of the use of simple indices described above is their simplicity in calculation and easy interpretation. However, the modern point processes statistics functions, which depend on the distances between all points of the pattern or distances between the nearest neighbors, are commonly used at present. Thus, functional summary statistics characterize a pattern as a function of scale. Depending on the data type, ecological questions to be answered and hypotheses to be tested, different functional summary statistics can be selected.

5.1. Nearest-neighbor distance-based distribution functions

There are a few functions that are able to quantify the spatial distribution of individuals as random, regular, or clumped. This is an important aspect of spatial structure of any population.

5.1.1. *Nearest-neighbor distance distribution function (G-function)*

The *G*-function is based on the distances from a point of the pattern (e.g., tree) to its nearest neighbor. The values of the *G*-function are nondecreasing as a function of distance *r*, starting from $G(r) = 0$. The nearest-neighbor distribution function for CSR is easy to calculate and it is equal to $G(r)=1-\exp(-\lambda \pi r^2)$. The empirical *G*-function is plotted against the theoretical expectation and it indicates how the individuals are spaced in the population. Clustered arrangement can be stated if $G_{obs} > G_{csr}$, and thus the nearest-neighbor distances between neighbors are smaller than it would be expected under randomness. In the case of regular pattern $G_{obs} < G_{csr}$, that is, the distances between nearest neighbors are larger than under random distribution [37, 68, 69].

5.1.2. *Empty-space function (F-function)*

The *F*-function characterizes the empty space in a pattern, and it is also known in the literature as the spherical contact distribution function. The function is based on the distribution of all distances between arbitrary selected points, but not the location of any point of the pattern, and its nearest neighbor [1]. The empty-space function characterizes the point pattern on the basis of the distances from the so-called test point to its nearest neighbors. This statistics is closely related to the *G*-function but its interpretation is opposite to that. The value of the *F*-function for CSR is the same as for *G*: $F(r)=1-\exp(-\lambda \pi r^2)$. The empirical *F*-function is again plotted against the theoretical values. Clumped distribution is assumed if the values of $F_{obs} > F_{csr}$. That is, the distances from an arbitrary point to its nearest neighbor of the pattern will be larger (on average) than under the CSR because the clustered pattern contains larger gaps than the random distribution. In the case of regular pattern, $F_{obs} > F_{csr}$, that is, the gaps are smaller and the distance from any point to its nearest neighbor will be smaller.

It is worth noting that both functions have their inhomogeneous versions, which can be applied in cases when the spatial pattern of individuals within the population is not homogeneous.

Example 5

Figure 10 presents the stem map generated from the data set collected in the 30-year old Scots pine (*Pinus sylvestris* L.) monoculture. The stand was planted artificially at the initial spacing 1.5 m × 1.5 m and it has not been managed so far. For each tree, the diameter at the breast height (dbh) was measured as well as location coordinates (*x*, *y*) were reported.

The nearest-neighbor distribution *G*-function (*G*) was calculated for the data, and the empirical function was plotted against the function for complete spatial randomness. Both functions are presented on the left panel in **Figure 11**. The graph of the *G*-function for the data set is clearly below the expectation indicating the regularity in trees distribution. Up to the distance of 1.8 m, $G(r) = 0$. This distance may be interpreted as the minimum distance between the nearest individuals and it is due to the hard-core process. This is the simplest kind of interaction between individuals.

http://dx.doi.org/10.5772/65320

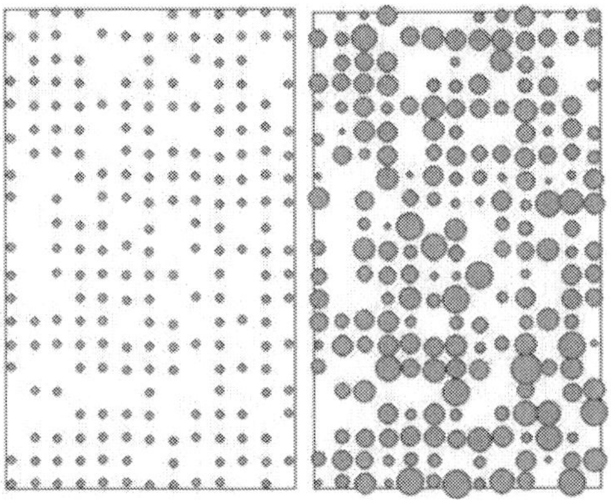

Figure 10. Stem map (left panel) of living trees in the 30-year old Scots pine (*Pinus sylvestris* L.) monoculture. In the right panel the size of circles corresponds to the diameter at the breast height.

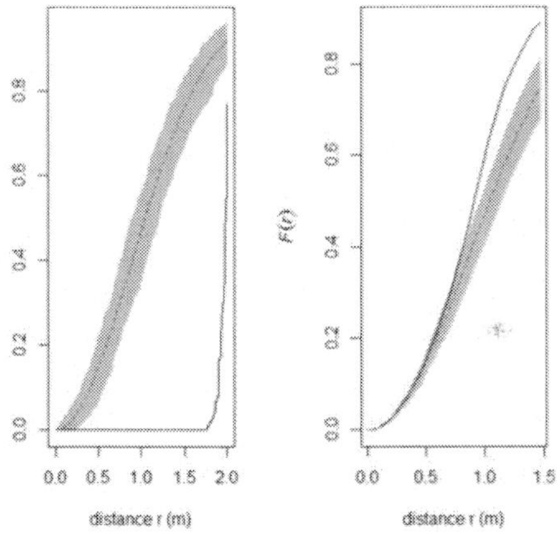

Figure 11. Nearest distance distribution *G*-function (left panel) and empty-space *F*-function (right panel) calculated for the data collected from the Scots pine monoculture. The solid line represents the empirical function for the data, the dashed line represents the function for CSR process, and the shadowed area represents the 95% pointwise confidence intervals calculated from 199 Monte Carlo simulations.

The empty-space *F*-function (*F*) is presented in the right panel in **Figure 11**. It confirms regularity in the spatial pattern of pines stated on the basis of the nearest-neighbor function.

Figure 12 presents the graphs of the *G*- and *F*-functions (left and right panels, respectively) calculated for hornbeams from an old-growth oak-dominated forest. Both functions confirmed the aggregated pattern of this tree species that is inconsistence with results obtained by spatial indices.

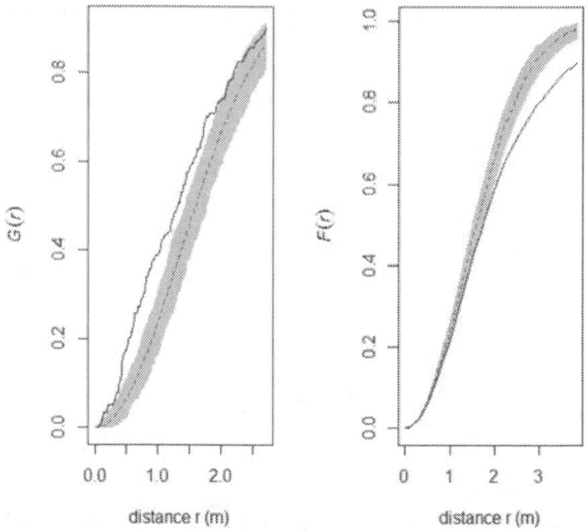

Figure 12. Nearest distance distribution *G*-function (left panel) and empty-space *F*-function (right panel) calculated for the data (hornbeam trees only) collected from an old-growth oak-dominated forest. Explanations: see **Figure 11**.

5.2. Second-order summary functions

Second-order statistics rely on the spatial relationships of pairs of trees, not only on nearest neighbor distances [37].

5.2.1. Second-order functions to discover the spatial arrangement of points

5.2.1.1. Univariate (unmarked) point pattern analysis

It refers to the pattern of points (e.g., trees in the forest) described only by their position (coordinates). Information on additional point attributes (e.g., size, sex, etc.) is not provided.

5.2.1.1.1. Ripley's function (K(r))

It appears to be the most common second-order summary function [1, 16, 44, 69]. This function is based on the measurements of distances between all individuals of the point pattern. It determines the expected number (λ) of points present within the distance r of typical point of the pattern. The expectation for the CSR is that there should be $\lambda\pi r^2$ individuals within the distance r of the typical point of the pattern. Under CSR the function yields $K(r) = \pi r^2$. For clustered pattern $K(r) > \pi r^2$ and for regular pattern $K(r) < \pi r^2$. Usually, the *K*-function is plotted —together with its expectation—against the different distances r (spatial scales). Its shape provides valuable information on the point pattern distribution. If the empirical $K(r) > \pi r^2$ it means that the distribution of the individuals within the population is consistent with clustering at the certain r distance. Opposite, the pattern is consistent with regularity if $K(r) < \pi r^2$. Because of the K-function increases at the rate of r^2 under the CSR expectation, it is better to use its transformation, the L-function, which stabilizes its variance and transforms $K(r)$ to the straight line $L(r) = r$ [37]. The interpretation of the L-function is quite easy. For regular distribution $L(r) < r$, and in the case of aggregated pattern—$L(r) > r$. To infer the scale of spatial interaction in a point pattern, it is obvious to estimate it by reading off the position where the function for the observed data set lies further away from the expectation under the CSR. It is not always correct because of its cumulative nature and effects at smaller distances obscure the effects at larger scales.

5.2.1.1.2. Pair correlation function ($g(r)$)

The alternative to the K-function is the pair correlation function, a noncumulative summary statistics. This function is closely related to the K-function and is recommended by [1, 8]. It contains the contributions only from interpoint distances equal to the distance r. The advantage of the $g(r)$-function is that under CSR it is equal 1 and independent of the intensity of the pattern. The tendency toward clustering means, that there will be more (on average) individuals at smaller distances r than expected under CSR and $g(r) > 1$. Conversely, for regular arrangement of individuals, there will be, on average, fewer individuals at the smaller distances than under CSR, and $g(r) < 1$ [1, 37].

5.2.1.2. Bivariate point pattern analysis

Both, Ripley's function and pair correlation function can be extended to discover spatial relationships between the points of two types. For example, bivariate point pattern analysis is a suitable tool to discover the spatial relationships between two different tree species mixed in the forest.

5.2.1.2.1. Bivariate Ripley's function ($K_{12}(r)$)

Ripley's function can be extended to the bivariate form and for more details on the suitable estimator, see Refs. [1, 8, 37]. The ecological questions here concern the detecting possible interactions between two types of objects (e.g., tree species in the forest). The fundamental benchmark is spatial independence separating two alternatives: association and repulsion (small scale) or segregation (large scale) of both types. Bivariate $L_{12}(r)$ is an analog of univariate $L(r)$-function. In case of the spatial independence of type 1 and type 2 of points $L_{12}(r) = r$. If $L_{12}(r) > r$ then two types of objects show spatial association at the certain distance r and if $L_{12}(r) < r$—points of different types show spatial repulsion (separation).

5.2.1.2.2. Bivariate pair correlation function ($g_{12}(r)$)

Similarly to the L-function, the $g(r)$-function can be easily extended to bivariate forms, $g_{12}(r)$, to discover correlations between two types of objects. Then, $g_{12}(r) = 1$ indicates the spatial independence of two types of points being at the distance r apart. If $g_{12}(r) > 1$ then spatial association of both types of objects can be stated and if $g_{12}(r) < 1$—they are spatially segregated at the distance r.

Both functions, Ripley's function and pair correlation function, can also be calculated for inhomogeneous point patterns, thus in the case of spatial variation in the intensity of the pattern [37].

Example 6

To present different shapes of univariate L- and g-functions for regular and aggregated patterns, data sets from Scots pine stand and old-growth oak-dominated forest, described previously, were used. Both functions for the empirical data sets are presented in **Figure 13**.

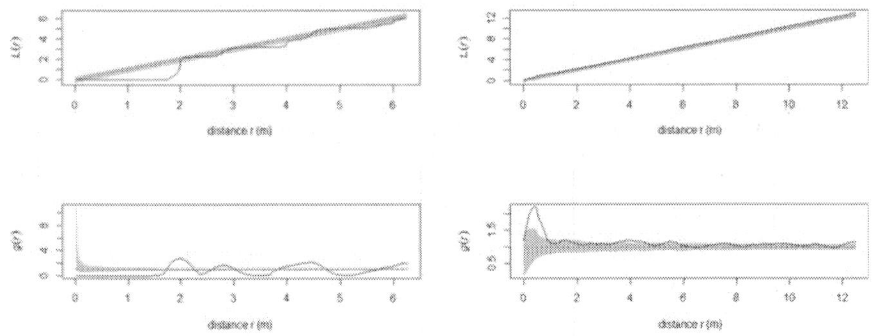

Figure 13. The L-function and g-function for 30-years old pine (*P. sylvestris* L.) stand (left panel) and hornbeam (*C. betulus* L.; right panel) in the old-growth oak-dominated forest. Dashed line represents the expected values of simulated pattern for complete spatial randomness (CSR), solid line represents empirical $g(r)$ function. Shaded area represents 95% pointwise confidence intervals based on 199 *Monte Carlo* simulations.

In the left panel, both functions calculated for live trees in pine stand showed clear evidence for regularity. Functions lie below the expectation referred to CSR and the departures from the expectation were significant at the distance up to 1.8 m (g-function) and 2 m (L-function). Up to these distances both functions are equal 0. It indicates the minimum distance between trees. Moreover, the shape of the pair correlation function is typical for plantations, where trees have been planted in rows that are also reflected by the wave-like shape of the function. Thus, the spatial pattern of trees can provide important information about the history of establishment of the forest.

In the right panel in **Figure 13**, there is an example of clustering of trees. Both functions lie above the expectation for CSR. Because the L-function has cumulative character it is rather hardly to make statements on the distance at which aggregations of trees can be observed. In the case of pair correlation function, this distance is clearly visible. The maximum value of the g-function at the certain distance is equal to the average cluster size of trees. In case of hornbeams it was about 0.5 m. Such small spatial aggregations of hornbeams are typical for regeneration from sprouts, which is quite frequently observed in the case of this tree species.

http://dx.doi.org/10.5772/65320

Spatial correlation between oak (subscript: db) and hornbeam (gb) —an example of bivariate analysis—is presented in **Figure 14**. Bivariate pair correlation function indicated spatial negative association (spatial repulsion) between these two tree species in the old-growth oak-dominated forest. It means that both trees are spatially separated. In virgin forests, spatial segregation is assumed to decrease the interspecific competition, and it is supported by different mechanisms, e.g., different niche requirements of tree species. Thanks to spatial separation of tree species they can coexist together in a multispecies forest.

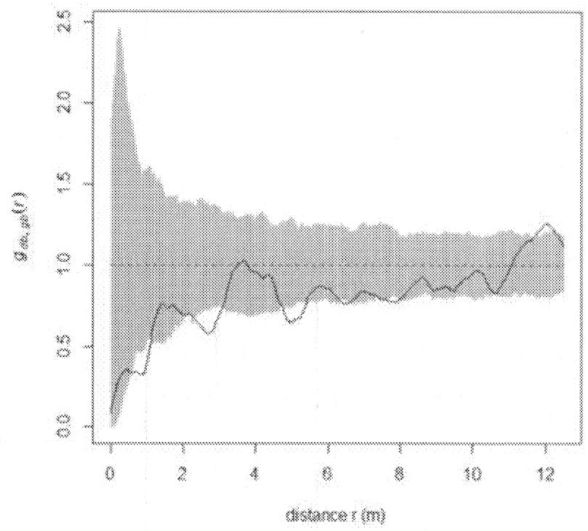

Figure 14. Bivariate pair correlation functions for oak (subscript: db) and hornbeam (gb) in an old-growth oak dominated forest. Solid line: empirical function; dashed line: expected value of the function for spatial independence of both species; shaded area: confidence region of null model acceptation calculated on the base of 199 Monte Carlo simulations.

In oak-dominated forest, the correlation range between oak and hornbeam was about $r = 11$ m, thus $g_{12}(r) < 1$ up to this distance. The negative association of both species results more likely from the extremely different abundance of oak and hornbeam as well as their different life stages. Clumped pattern of hornbeam may results from sprouting while random distribution of oak is typical for old, large trees. In plant populations, low intraspecific competition and higher interspecific competition favor species coexistence in multispecies forests.

5.2.2. Inhomogeneous point pattern analysis

Inhomogeneous point pattern analysis should be used in cases when point density differs significantly with their location. Such cases are frequently observed in the natural forests, e.g., due to the forest site variation, seed dispersion, etc. Incorrect use of the second-order summary function leads to misinterpretation of the results, the so-called virtual aggregation. To avoid it, one can use inhomogeneous versions of the summary functions mentioned above or special function introduced by Schiffers et al. [70].

5.2.2.1. *K2-function*

This function was developed as an extension of $g(r)$ that can be used to discover the regular or clumped patterns despite the presence of the spatial variation in the point intensity across the study region [70]. Unlike the *L*- and *g*-functions, the *K*2-function relates the intensities at a given scale to the intensities at the adjacent scales [70]. It allows to interpret scales of significant deviations from the expectations at distances where transitions from low (or high) to high (low) intensities occur. The negative values of the *K*2-function indicate clustering because the neighborhood density decreases with increasing distance. It has the positive values for regular pattern due to the steep increase of neighborhood density at a certain distance.

Example 7

Figure 15 presents stem map generated for European yew (*Taxus baccata* L.) located in the Kórnik Arboretum, western Poland [71]. The population of yew developed spontaneously during last decades. The map represents the location of male individuals only.

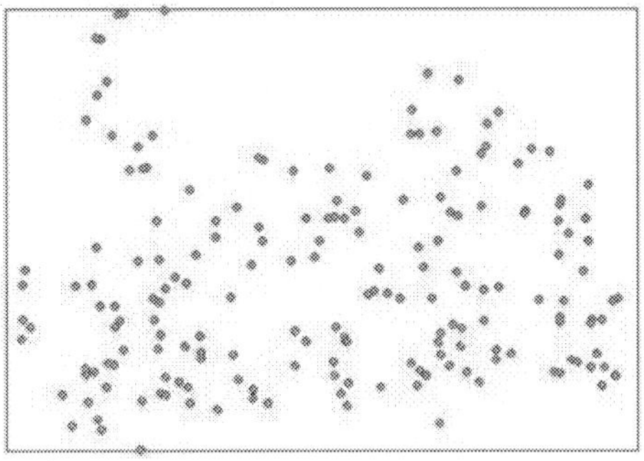

Figure 15. Stem map for males of yew (*Taxus baccata* L.). Trees are represented by points irrespectively of their diameter.

Visual inspection provides information that the density of males across the study plot was inhomogeneous, and there is a density gradient from the south (bottom) to the north (top) of the plot. Inhomogeneity in the tree density can be clearly seen on the graph with pair correlation function that lies completely above the value 1 indicating the so-called virtual aggregation due to the heterogeneity in tree density because the pair correlation function is related to the global intensity in the surrounding of a tree.

Thus, pair correlation function would lead to misinterpretation about the aggregated structure of males. The dependence in global intensity restriction is circumvented by the *K*2-function. In the right panel of **Figure 16**, the estimated *K*2-function lies completely within the confidence region under the CSR expectation. There are only weak deviations (statistically insignificant) at the smallest spatial scale toward clumping of males. Thus, the distribution of males did not differ from the randomness.

http://dx.doi.org/10.5772/65320

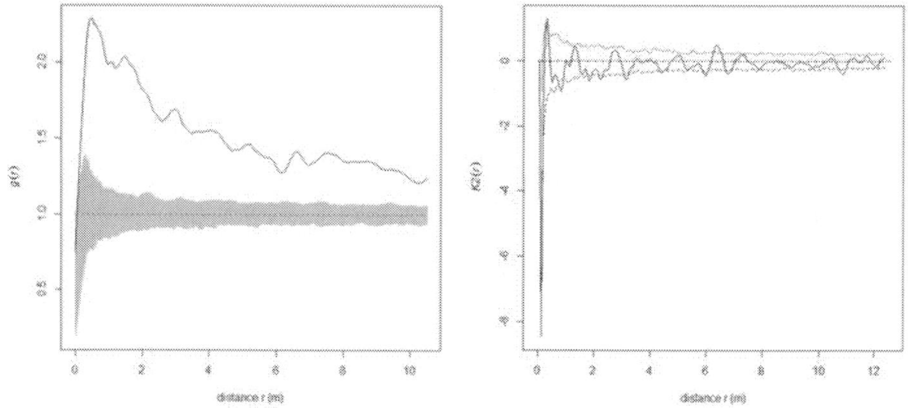

Figure 16. Pair correlation function (left panel) and *K*2-function (right panel) calculated for males of European yew (*T. baccata* L.). Solid lines represent empirical functions and dashed line represents expectation under the CSR process. Dashed region (left panel) and dotted lines (right panel) represent confidence region of the null model (CSR) acceptation calculated on the base of 199 Monte Carlo simulations.

5.2.3. Marked point pattern analysis: spatial diversity of different plant attributes

Marked point pattern carries different marks (attributes) of points. Marks can be qualitative and quantitative. In this section, methods suitable to analyze the correlations among plant's attributes. (e.g., sizes, health status, etc.) are provided with real data examples.

5.2.3.1. Qualitative marks

Marked point pattern analysis for qualitative marks describes the points in a different way than in the case of bivariate pattern analysis (like in Section 5.2.1.2). Here, the mark is produced by the process acting a posteriori over the univariate pattern, and it is a fundamental difference to the bivariate pattern in which plant's attributes are generated a priori by two different processes (e.g., plant species) [72]. It means that qualitative marks are defined as something created conditional on a given pattern [1].

5.2.3.1.1. Mark connection function (p_{12})

This function is the conditional probability, given that there is a point of the process at the location m and the second point at the location n and they are separated by the distance r such that the first individual is of type 1 and the second one is of type 2 [8, 37]. If the marks attached to the points (e.g., trees) of the pattern are independent and identically distributed, then $p_{12}(r) = p_1p_2$, where p_1 and p_2 denote the probability that a point is of type 1 or 2, respectively. Values larger than this, $p_{12}(r) > p_1p_2$, indicate positive association between the two types, while $p_{12}(r) < p_1p_2$ indicates the negative association.

Example 8

The mark connection function was applied to test whether there was any spatial correlation between trees of different health status of European yew (*T. baccata* L.). The study plot (**Figure 17**) was established in the Kniazdwor Nature Reserve, western Ukraine [73]. Yew occurred under the canopy of European beech (*Fagus sylvatica* L.) and silver fir (*Abies alba* L.). All individuals of the height $h > 0.5$ m were classified according to the simple general classification: 1, good health status; 2, poor health status. Details on the classification can be found in Ref. [71].

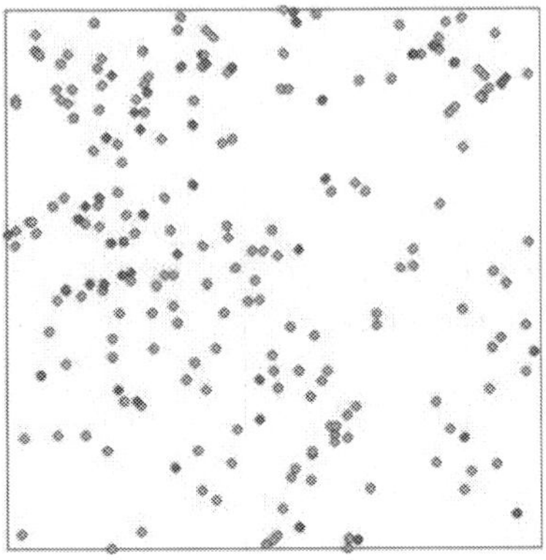

Figure 17. Stem map generated from the data set collected in the Kniazdwor Nature Reserve, Ukraine. Points represent yew (*T. baccata* L.) trees with different health status. Green dots: yews of good health status; red dots: yews of poor health status.

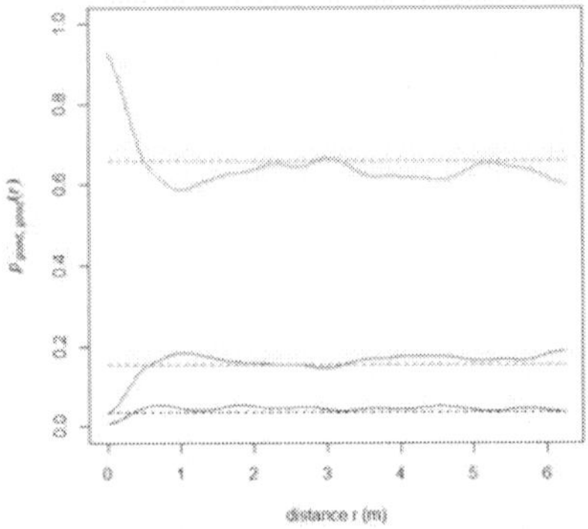

Figure 18. Mark connection function for health status—good vs. poor—of yew trees in the Kniazdwor Nature Reserve, Ukraine. Dashed lines: the reference values of $p_{12}(r)$-function (red), $p_{11}(r)$ (green), and p_{22}(black) for random allocations of marks; solid line: estimated $p_{12}(r)$-function for the data set.

Trees of poor health status showed neither the negative nor the positive association, that is, the function $p_{22}(r) \approx p_2 p_2$ (black solid line, **Figure 18**). Because trees of good health status showed highly clustered structure at small spatial scale, the probability of finding two healthy trees close to each other was higher than expected ($p_{11}(r) > p_1 p_1$). Healthy trees have—over the same spatial scale—a lower than expected probability of having trees of poor health status as its neighbor, that is, $p_{12}(r) < p_1 p_2$ (**Figure 18**).

Healthy tree have—over the same spatial scale—a lower than expected probability of having trees of poor health status as its neighbor, that is, $p_{12}(r) < p_1 p_2$ (**Figure 18**).

5.2.3.2. Quantitative marks

Quantitative marks additionally describe each tree and they are numerical values (e.g., stem diameter, tree height, etc.). One can be interested in finding out whether the sizes of trees growing at the distance r from each other show any spatial correlation, conditional on the their location (unmarked pattern). An appropriate summary statistics for quantitative data types are different mark correlation functions depending on the so-called test function used in calculation [1, 7, 14, 16, 40, 74]. Two correlation functions seem to be especially important in the structural analysis of the population: mark correlation function and mark variogram.

5.2.3.2.1. Mark correlation function ($k_f(r)$)

This function is a measure if the dependence between marks of two individuals of the pattern is separated by the distance r [8]. The test function with two marks, m_1 and m_2, is a nonnegative number and the test function is of the following form: $t(m_1, m_2) = m_1 m_2$. The normalized k_f for a random assignment of marks (lack of spatial correlation among marks) over the pattern is equal to 1. Values of $k_f(r) < 1$ for the distance r mean that both individuals have smaller marks than the average for the population. At the small distances it means that there is an inhibition between both individuals due to their close distance. If $k_f(r) > 1$ it means that two individuals growing at the distance r show larger marks than the average. At small distances it means that they benefit from being close to one another [8]. Moreover, it offers another characteristic of the pattern, namely, correlation range. It is the distance r at which the function approaches the value 1. Using this form of correlation function, one is interested in finding out whether the marks of two plants show any correlation in space.

5.2.3.2.2. Mark variogram ($\gamma(r)$)

In this form of correlation function the test function is $t(m_1, m_2) = 0.5(m_1 - m_2)^2$. It characterizes the squared differences between marks of pairs of individuals with the distances of r. If individuals growing at the distance r apart have similar mark, then mark variogram has smaller (than under random condition) values. Large values of $\gamma(r)$ indicate that marks of both individuals tend to be different at a certain distance r. Similarly to the $k_f(r)$ function, the correlation range can be stated [8, 37, 40].

Example 9

Figure 19 presents the mark correlation function for diameter of trees in the oak-dominated old-growth forest from Example 1. Analysis of $k_f(r)$ indicated that pairs of trees growing at the distance up to 9 m (correlation range) tended to have smaller diameters than the average for the stand. In ecological meaning it can be interpreted as the mutual growth inhibition of the neighboring trees. Mark variogram showed another interesting point of view providing details on similarity of dissimilarity of pairs of trees in the dependence on the distance r between them. In the oak-dominated forest, live trees being close to one another tended to have similar diameters and the interaction range was about 12 m.

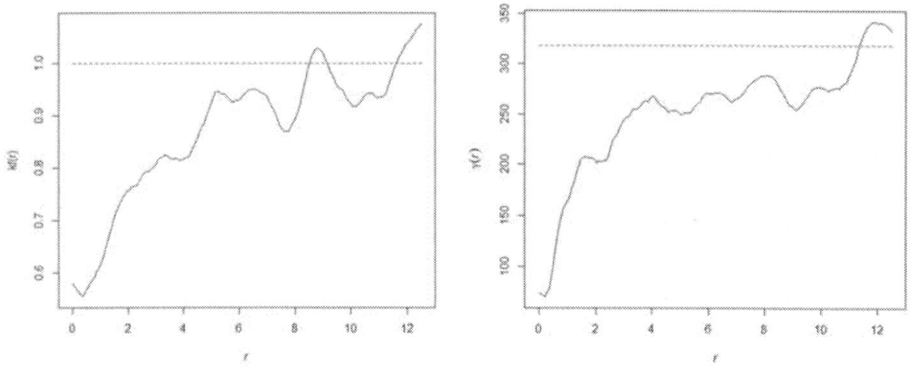

Figure 19. Mark correlation function (left panel) and mark variogram (right panel) for diameters of live trees in the old-growth oak-dominated forest. Dashed line represents the function for random allocation of marks of trees meaning their lack of spatial correlation.

6. Conclusion

Spatial analyses have now largely been incorporated in ecological studies due to the realistic assumption of spatial dependence between individuals constituting plant populations. Population structure is one of the most important traits of each biosystem and its description allows deeper insight into mechanisms and processes responsible for population dynamics. To understand these natural processes, modifying the structure and dynamics of plant populations is important from ecological (scientific) and practical (managing of natural resources) point of view. As indicated, depending on the ecological questions stated, different methods of spatial point pattern analysis can be applied. All of them are suitable to extract the hidden and detailed information on the current state of any population and allow us to make the assumptions concerning their future development. It is important to remember that the key elements of spatial analysis in ecology are data type, the appropriate choice of summary statistics, and null models. Selecting few of them in a single analysis makes the statements more reliable and realistic in the changing world.

http://dx.doi.org/10.5772/65320

Author details

Janusz Szmyt

Address all correspondence to: jszmyt@up.poznan.pl

Department of Silviculture, Faculty of Forestry, Poznań University of Life Sciences, Poznań, Poland

References

[1] Wiegand T, Moloney AK: Handbook of Spatial Point-Pattern Analysis in Ecology. Boca Raton, FL: CRC Press, Taylor & Francis Group; 2014. 510 pp.

[2] Liebhold AM, Gurevitch J: Integrating the statistical analysis of spatial data in ecology. Ecography. 2002; 25: 553–557.

[3] Szmyt J: Spatial statistics in ecological analysis: from indices to functions. Silva Fennica. 2014; 48: e1–e33.

[4] Kenkel NC: Pattern of self-thinning in jack pine testing the random mortality hypothesis. Ecology. 1988; 9: 1017–1024.

[5] Lepš J: Can underlying mechanisms be deduced from observed patterns? In: Krahulec F, Agnew ADQ, Agnew S, Willems JH, editors. Spatial Processes in Plant Communities. The Hague: APB Academic Publication; 1990. pp. 1–11.

[6] McIntire E, Fajardo A: Beyond description: the active and effective way to infer processes from spatial patterns. Ecology. 2009; 90 (1): 46–56.

[7] Raventós J, Wiegand T, Luis M: Evidence for the spatial segregation hypothesis: a test with nine-year survivorship data in a Mediterranean shrubland. Ecology. 2010; 91 (7): 2110–2120.

[8] Illian JB, Moller J, Stoyan H, Stoyan: Statistical Analysis and Modelling of Spatial Point Patterns. Chichester: John Wiley & Sons; 2008. 510 pp.

[9] Pretzsch H: Forest Dynamics, Growth and Yield. Heidelberg: Springer; 2010. 664 pp.

[10] Pommerening A: Approaches to quantifying forest structures. Forestry. 2002; 75 (3): 305–324.

[11] Comas C, Palahi M, Pukkala T, Mateu J: Characterizing forest spatial structures through inhomogenous second order characteristics. Stochastic Environmental Research and Risk Assessment. 2009; 23: 387–397.

[12] del Río M, Pretzsch H, Alberdi I, Bielak K, Bravo F, Brunner A: Characterization of the structure, dynamics, and productivity of mixed-species stands: review and perspectives. European Journal of Forest Research. 2015; 135(1):23–49. doi: 10.1007/s10342-015-0927-6

[13] Perry JN, Liebhold AM, Rosenberg MS, Dungan M, Miriti J, Jakomulska A, Citron-Pusty S: Illustrations and quidelines for selecting statistical methods for quantifying spatial pattern in ecological data. Ecography. 2002; 25: 278–600.

[14] Pommerening A, Goncalves A, Rodriguez-Soalleiro R: Species mingling and diameter differentiation as second-order characteristics. Allgemeine Forst und Jagdzeitung. 2011; 182: 115–129.

[15] Penttinen A: Statistics for marked point patterns. In Yearbook of the Finnish Statistical Society; The Finnish Statistical Society; 2007. pp. 70–91; Helsinki, Finland.

[16] Velázquez E, Paine CET, May F, Wiegand T: Linking trait similarity to interspecific spatial associations in a moist tropical forest. Journal of Vegetation Science. 2015; 26: 1068–1079. doi: 10.1111/jvs.12313

[17] Shimatani K, Kubota Y: Spatial analysis for continuously changing point patterns along a gradient and its application to an *Abies sachalinensis* population. Ecological Modelling. 2004; 180: 359–369. doi: 10.1016/j.ecolmodel.2004.04.036

[18] Picard N, Bar-Hen A, Mortier F, Chadoeuf J: Understanding the dynamics of an undisturbed tropical rain forest from the spatial pattern of trees. Journal of Ecology. 2009; 97: 97–108. doi: 10.1111/j.1365-2745.2008.01445.x

[19] Martínez I, Wiegand T, González-Taboada F, Obeso JR: Spatial associations among tree species in a temperate forest community in North-western Spain. Forest Ecology and Management. 2010; 260: 456–465. doi: 10.1016/j.foreco.2010.04.039

[20] Zhang C, Zhao X, Gao L, von Gadow K: Gender-related distribution of Fraxinus mandshurica in secondary and old-growth forests. Acta Occologica. 2010; 36: 55–62. doi: 10.1016/j.actao.2009.10.001

[21] Wang X, Wiegand T, Wolf A, Howe R, Davies SJ, Hao Z: Spatial patterns of tree species richness in two temperate forests. Journal of Ecology. 2011; 99: 1382–1393. doi: 10.1111/j.1365-2745.2011.01857.x

[22] Wehenkel C, Brazao-Protazio JM, Carrillo-Parra A, Martinez-Guerrero HJ, Crecente-Campo F: Spoatial distribution patterns in the very rare and species-rich Picea chihuahuana tree community (Mexico). PLOS One. 2015; 10: e0140442. doi: 10.1371/journal.pone.0140442

[23] Wu LC, Li HQ: Summary statistics for measuring the relationship among three types of points in multivariate point patterns. Computational Statistics and Data Analysis. 2009; 53: 2809–2816. doi: 10.1016/j.csda.2008.10.009

[24] Grabarnik P, Myllymäki M, Stoyan D: Correct testing of mark independence for marked point patterns. Ecological Modelling. 2011; 222: 3888–3894. doi: 10.1016/j.ecolmodel. 2011.10.005

[25] Szmyt J, Stoyan D: Spatial autocorrelation of tree attributes in naturally regenerated managed beech (Fagus Sylvatica) forests in the Beskid Niski Mountains, southern Poland. Dendrobiology. 2014; 71: 129–136.

[26] Spies T: Forest structure: a key to the ecosystem. Northwest Science. 1998; 72: 34–39.

[27] von Gadow K, Zhang CY, Wehenkel C, Pommerening A, Corral-Rivas J, Korol M, Myklush S, Hui HY, Kiviste A, Zhao XH: Forest structure and diversity. In: Pukkala T, von Gadow K, editors. Continuous Cover Forestry. Managing Forest Ecosystems. Dordrecht: Springer; 2012. 296 pp. doi: 10.1007/978-94-007-2202-6_2

[28] Fortin M, Dale M, Hoef JV: Spatial analysis in ecology. In: El-Sharaawi AH, Piegorsch WW, editors. Encyclopedia of Environmetrics. Chichester: John Wiley & Sons; 2002. pp. 2051–2058.

[29] Stoll P, Prati D: Intraspecific aggregation alters competitive interactions in experimental plant communities. Ecology. 2001; 82: 319–327. doi: 10.2307/2679862

[30] Perry GLW, Miller BP, Enright NJ: A comparison of methods for the statistical analysis of spatial point patterns in plant ecology. Plant Ecology. 2006; 187 (1): 59–82. doi: 10.1007/s11258-006-9133-4

[31] Miller BP, Perry GLW, Enright NJ, Lamont BB: Contrasting spatial pattern and pattern-forming processes in natural vs. restored shrublands. Journal of Applied Ecology. 2010; 47: 701–709. doi: 10.1111/j.1365-2664.2010.01816.x

[32] Fajardo A, McIntire EJB: Distinguishing microsite and competition processes in tree growth dynamics: an a priori spatial modeling approach. American Naturalist. 2007; 169 (5): 647–661. doi: 10.1086/513492

[33] Wiegand T, He F, Hubbell SP: A systematic comparison of summary characteristics for quantifying point patterns in ecology. Ecography. 2013; 36 (1): 92–103. doi: 10.1111/j. 1600-0587.2012.07361.x

[34] Murrell DJ, Purves DW, Law R: Uniting pattern and process in plant ecology. Trends in Ecology and Evolution. 2001; 16: 529–530.

[35] Flügge AJ, Olhede SC, Murrell DJ: A method to detect subcommunities from multivariate spatial associations. Methods in Ecology and Evolution. 2014; 5 (11): 1214–1224. doi: 10.1111/2041-210X.12295

[36] Myllymäki M, Särkkä A, Vehtari A: Hierarchical second-order analysis of replicated spatial point patterns with non-spatial covariates. Spatial Statistics. 2014; 8: 104–121. doi: 10.1016/j.spasta.2013.07.006

[37] Baddeley A, Rubak E, Turner R: Spatial Point Patterns. Methodology and Applications with R. Boca Raton, FL: CRC Press; 2016. 810 pp.

[38] Stoll P, Bergius E: Pattern and process: competition causes regular spacing of individuals within plant populations. Journal of Ecology. 2005; 93: 395–403. doi: 10.1111/j.1365-2745.2005.00989.x

[39] Gray L, He F: Spatial point-pattern analysis for detecting density-dependent competition in a boreal chronosequence of Alberta. Forest Ecology and Management. 2009; 259: 98–106. doi: 10.1016/j.foreco.2009.09.048

[40] Pommerening A, Särkkä A: What mark variograms tell about spatial plant interactions. Ecological Modelling. 2013; 251: 64–72.

[41] LeMay V, Pommerening A, Marshall P: Spatio-temporal structure of multi-storied, multi-aged interior Douglas fir (Pseudotsuga menziesii var. glauca) stands. Journal of Ecology. 2009; 97: 1062–1074. doi: 10.1111/j.1365-2745.2009.01542.x

[42] Fibich P, Leps J, Novotny V, Klimes P, Tesitel J, Molem K, Damas K, Weiblen GD: Spatial patterns of tree species distribution in New Guinea primary and secondary lowland rain forest. Journal of Vegetation Science. 2016; 27: 1–12. doi: 10.1111/jvs.12363 10.1111/jvs.12363

[43] Krebs CJ: Ecological Methodology. 2nd ed. Addison-Welsey Longman; 1999. 620 pp.

[44] Ripley BD: Spatial Statistics. John Wiley & Sons; 2004. 525 p.

[45] Reich RM, Davis R: Quantitative Spatial Analysis. Fort Collins: Colorado State University; 2008. 511 pp.

[46] Szewczyk J, Szwagrzyk J: Spatial and temporal variability of natural regeneration in a temperate old-growth forest. Annals of Forest Science. 2010; 67: 202p1–202p8. doi: 10.1051/forest/2009095

[47] Rosenberg MS, Anderson CD: PASSaGE: pattern analysis, spatial statistics and geographic exegesis. Version 2. Methods in Ecology and Evolution. 2011; 2: 229–232. doi: 10.1111/j.2041-210X.2010.00081.x. DOI:10.1111/j.2041-210X.2010.00081.x#_blank

[48] Habashi H, Hosseiniand S: Stand structure and spatial patterns of trees in mixed Hyrcanian Beech forest, Iran. Pakistan Journal of Biological Sciences. 2007; 10: 1205–1212.

[49] Sankey TT: Spatial patterns of Douglas-fir and aspen forest expansion. New Forests. 2007; 35: 45–55.

[50] Sapkota IP, Tigabu M, Odén PC: Spatial distribution, advanced regeneration and stand structure of Nepalese Sal (Shorea robusta) forests subject to disturbances of different intensities. Forest Ecology and Management. 2009; 257: 1966–1975. doi: 10.1016/j.foreco.2009.02.008

[51] Amanzadeh B, Sagheb-Talebi K, Foumani B, Fadaie F, Camarero J, Linares J: Spatial distribution and volume of dead wood in unmanaged Caspian Beech (Fagus orientalis) forests from northern Iran. Forests. 2013; 26: 751–65. doi: 10.3390/f4040751

[52] Diggle PJ: Statistical Analysis of Spatial and Spatio-Temporal Point Patterns. 3rd ed. Boca Raton, FL: Taylor & Francis; 2014. 267 pp.

[53] Staupendahl K, Zucchini W: Estimating the spatial distribution in forest stands by counting small angles between nearest neighbours. Allgemeine Forst und Jagdzeitung. 2006; 177: 160–168.

[54] Donnelly K: Simulations to determine the variance and edge-effect of total nearest neighbor distance. In: Hodder I, editor. Simulation Methods in Archeology. London: Cambridge University Press; 1978. pp. 91–95.

[55] Kint V, Robert DW, Noël L: Evaluation of sampling methods for the estimation of structural indices in forest stands. Ecological Modelling. 2004; 180: 461–476. doi 10.1016/j.ecolmodel.2004.04.032

[56] Neumann M, Starlinger F: The significance of different indices for stand structure and diversity in forests. Forest Ecology and Management. 2001; 145: 91–106.

[57] Szmyt J: Spatial structure of managed beech-dominated forest: applicability of nearest neighbors indices. Dendrobiology. 2012; 68: 69–76.

[58] Kint V, Van Meirvenne M, Nachtergale L, Geudens G, Lust N: Spatial methods for quantifying forest structure development: a comparison between nearest-neighbor indices and variogram analysis. Forest Science. 2003; 49: 36–49.

[59] Pukkala T, von Gadow K: Forest structure and diversity. In: Pukkala T, editor. Continuous Cover Forestry. Springer Science + Business Media B.V.; 2012. 296 pp.

[60] Corral#x2010;Rivas JJ, Pommerening A, Gadow K von, Stoyan D: An analysis of two directional indices for characterizing the spatial distribution of forest trees. In: Models of tree growth and spatial structure for multi-species, uneven-aged forests in Durango (Mexico). PhD dissertation. 2006. Gottingen, Faculty of Forest Science and Forest Ecology, Georg#x2010;August University of Gottingen. 106–121 p.

[61] Aguirre O, Hui G, von Gadow K, Jiménez J: An analysis of spatial forest structure using neighbourhood-based variables. Forest Ecology and Management. 2003; 183: 137–145.

[62] Vorčak J, Merganic J, Saniga M: Structural diversity change and regeneration processes of the Norway spruce natural forest in Babia hora NNR in relation to altitude. Journal of Forest Science. 2006; 52: 399–409.

[63] Corral-Rivas JJ, Wehenkel C, Castellanos-Bocaz HA, Vargas-Larreta B, Diéguez-Aranda U: A permutation test of spatial randomness: application to nearest neighbour indices in forest stands. Journal of Forest Research. 2010; 15 (4): 218–225. doi: 10.1007/s10310-010-0181-1

[64] Hui G, Zhao X, Zhao Z, von Gadow K: Evaluating tree species spatial diversity based on neighborhood relationships. Forest Science. 2011; 57: 292–300.

[65] Pielou EC: Mathematical Ecology. New York: John Wiley & Sons; 1977. 385 pp.

[66] Graz FP: The behavior of the species mingling index M_{sp} in relation to species dominance and dispersion. European Journal of Forest Research. 2004; 123: 87–92. doi: 10.1007/s10342-004-0016-8

[67] Motz K, Sterba H, Pommerening A: Sampling measures of tree diversity. Forest Ecology and Management. 2010; 260: 1985–1996. doi: 10.1016/j.foreco.2010.08.046

[68] Mateu J, Uso J, Montes F: The spatial pattern of a forest ecosystem. Ecological Modelling. 1998; 108: 163–174.

[69] Dixon PM: Nearest neighbor methods. In: El-Sharaawi AH, Piegorsch WW, editors. Encyclopedia of Environmetrics. Chichester: John Wiley & Sons; 2001. pp. 1–26.

[70] Schiffers K, Schurr FM, Tielbörger K, Urbach C, Moloney K, Jeltsch F: Dealing with virtual aggregation – a new index for analysing heterogeneous point patterns. Ecography. 2008; 31: 545–555. doi: 10.1111/j.0906-7590.2008.05374.x

[71] Iszkuło G, Boratyński A: Interaction between canopy tree species and European yew Taxus baccata (Taxaceae). Polish Journal of Ecology. 2004; 52: 523–531.

[72] Goreaud F, Pélissier R: Avoiding misinterpretation of biotic interactions with the intertype K12-function: population independence vs. random labelling hypotheses. Journal of Vegetation Science. 2003; 14: 681–692.

[73] Iszkuło G, Boratyński A: Different age and spatial structure of two spontaneous subpopulations of Taxus baccata as a result of various intensity of colonization process. Flora. 2005; 200: 195–206.

[74] Wälder K, Wälder O: Analysing interaction effects using the mark correlation functions. iForest. 2007; 1: 34–38.

Chapter 6

Practical Value of User-Centred Spatial Statistics for Responsive Urban Planning

Damjan Marušić and Barbara Goličnik Marušić

Additional information is available at the end of the chapter

http://dx.doi.org/10.5772/65322

Abstract

This chapter addresses spatial statistics via an alternative perspective, focusing on evidence-based people-spatial relationships and related measures, quantifications and qualifications, and by this, it provides rather specific spatial information and spatial statistics about urban environments. It is based on time quality assessment (TQA), a time-people-place-oriented approach for the analysis and simulation of the quality of living environments, backgrounded with the method of behaviour mapping. It shows that the quality of the time spent on a certain activity in a certain place indicates the quality of the living environment. It also shows that the quality of the time spent depends on what a person can afford, and it provides an evaluation of the quality of living environments with a measure of good/bad time. The practical value is in the provision of empirical knowledge to support planning guidance based on user-centred small-scale spatial statistics, which is able to inform top-down and bottom-up decision-making processes for people-friendly living environments.

Keywords: spatial-temporal statistics, urban planning, quality of life, behaviour mapping, bottom-up, user-centred, evidence-based, time quality

1. Introduction

Our (urban) living environment, composed of material and non-material components and relations among them, including an infrastructure and other built components, ecosystems, their inhabitants and users (e.g. people, animals, vegetation) and other entities (e.g. various enterprises, cultural and political entities, etc.), is a dynamic, complex system (e.g. [1–3]). In general, such a system is unpredictable (e.g. [4, 5]). It is composed of known invariable components (e.g. macro-location, general climate conditions, certain elements of the environment in the considered time period, etc.); known variables (those of which we are aware, but their quality or quantity is unknown or variable, e.g. infrastructure, [built] environment, individuals, their habits, their occupations, their routines, etc.); and unknown variables (those of whom we are not aware and/or cannot determine, e.g. daily politics and unpredictable disasters).

In approaching the system from a large scale in certain circumstances, the analysis and simulation of selected aspects within the selected condition limits may appear simple (i.e. it is possible to determine the simple approximations of relations, e.g. linear). However, such an approach in general does not result in appropriate useful results. On the other hand, the determination of very complex relations in the large-scale analyses may result in very unreliable results and uncontrollable simulations.

Considering the above, an attempt to start with a more profound investigation of the components of the system that are of concern to us and their relations to other components, i.e. to the system, appears reasonable. At that level, relations are more simple (or less complex), and it is more likely that simple approximations result in acceptable outcomes. Yet, the simultaneous monitoring and valuation of higher-level results (i.e. mass result, larger scale) is important. This calls for the use of the bottom-up approach, based on small rather than big data, which may help with interpreting or informing big data in spatial statistics (e.g. [6]).

This chapter addresses people-spatial relationships, their quantifications and qualifications and related measures for bottom-up evidence-based and user-centred urban planning. Based on time quality assessment (TQA), a time-people-place-oriented approach for the evaluation of the quality of living environments, this chapter provides specific types of spatial information about urban environments and challenges the current two-dimensional land-use perspective in urban planning with a dynamic and comprehensive perspective, taking into account users, the activities in which they are involved and the environments in which these activities are taking place, analysing three key parameters: time balance, economic balance and time-quality balance. The chapter shows that the quality of the time spent on a certain activity in a certain place indicates the quality of the living environment. It also shows that the quality of the time spent depends on what a person can afford, and it provides an evaluation of the quality of living environments with a measure of good/bad time. This chapter provides empirical knowledge based on user-centred small-scale spatial statistics to support practical guidance for informing decision-making processes for people-friendly living environments.

In relation to spatial planning assistance, keeping behaviour patterns in mind, interest has been increasing in the development and implementation of approaches based on information computer technology (ICT) and geolocated social media data (e.g. [7, 8]). However, to be able to talk about the quality of living environments via a time-people-place-oriented approach, first, time as a dimension also has to be applied to non-transportation activities. Reference [9] addressed the travel-time ratio and examined the relationship between travel time and stay time (keeping the goal of the travel in mind). Such an approach is particularly useful in the evaluation of the actual temporal scheme of a person's routine. However, it does not comment on the quality of the time spent on travelling or on staying. In this respect, the chapter addresses the quality of living (environments) via the quality of time spent within peoples' daily routines.

The quality of time spent on an activity is a complex function going beyond the quantity of time spent on a certain activity in a certain place. However, it combines the basic economic ability of a profile, the assessment of the conduciveness of the physical environment and the pleasantness of the activity taking place there. Contemporary responsive urban planning on a general level refers to the quality of living environments and well-being. In relation to the development and current state of the field of approaches towards assessing or measuring quality of living, a variety of comprehensive concepts related to quality of life exist, usually referring to the quantitative social, spatial and economic aspects (e.g. [10–13]). A literature review shows that although quality of life is recognised as a general concern, little consensus has been reached on a definition of quality of life or on the factors/predictors of an individual's quality of life (e.g. [14, 15]). In the past decade, the quality-of-life concept has also been focussing on well-being, health and standard of living addressed via softer indicators, such as happiness, life satisfaction and the like [16].

However, despite the fact that many strategic documents (e.g. [17]) presenting fundamental objectives for smart, sustainable and inclusive growth have emphasised the importance of local development towards quality of place and the well-being of people, quality-of-life-oriented studies still lack a focus on detailed actual and local-level aspects, which may better interpret or indicate quality of life and living environments. In relation to this, [18] argues that the actual implementation of such objectives in real-life situations (on a scale of 1:1) is often vaguely realised. In this framework, this chapter introduces the prototype of the TQA approach and shows how the model can work. TQA has been introduced as an alternative approach for assessing the effectiveness of human environments for living [19], using bottom-up evidence-based spatial statistics. In city planning and design processes, the TQA of living environments represents a potential universal baseline, where the TQA approach examines relationships between characteristic socioeconomic profiles acting in certain environments.

2. TQA approach

The current development stage of the TQA approach is characteristic of a clearly stated and well-developed concept, based mostly on theoretical simulations. A fully fledged investigation was not implemented. The concept foresees that calibration regarding quality of activity follows target-group questionnaires, interviews or appropriate ways of crowd sourcing (e.g. Web public participation, social networks) depending on the environment where the approach is applied. Similarly, quality parameters and weights used initially follow a combination of expert knowledge (e.g. sociological studies of everyday life, studies addressing placemaking and place attachment, a combination of expert knowledge from the fields of environmental psychology, urban planning and design) and data collected from the relevant target groups. This chapter discusses a new approach and illustrates its applicability to and value mostly for examples that simulate possible real situations. The comments are based on selected cases, theoretically set up and occasionally proven for some territories, knowing their socioeconomic characteristics (source: Statistical Office of the Republic of Slovenia [SURS]; Surveying and Mapping Authority of the Republic of Slovenia [GURS]), place characteristics (e.g. spatial-site analysis, behaviour-mapping analysis, GURS) and commuting possibilities for the theoretical target profile, using Michelin or similar portals. To keep the discussion manageable, the simplification of parameters or situations is used.

Three main pillars of input data are relevant for the approach: data related to the user profile, data related to the activity for which the suitability of the area is examined and data related to space. In general, the collected data refer to five binds of information: population, housing, leisure and recreation, services and transport, and they provide a possible examination of:

- facilities that examined territories shall provide,
- mobility networks that assure accessibility to these facilities and
- facilities in correlation with population densities.

The population can be grouped into various groups, based on common crucial characteristics, resulting in segments of the population. One such segment of the population is defined by boundary profiles and characteristic profiles (e.g. central profile, the most representative). Accordingly, it is possible to define the limits of the population of the studied area and the edge conditions of/for such a population within the area. Further, individual profiles are necessary to define, as they can help to describe the population in the studied area. They can be set up from available statistical data or any other relevant source (e.g. questionnaire) regarding demographic and social parameters, such as age, gender, family status, education, occupation, income and the like.

Based on crucial boundary characteristics, variations of individual profiles are designed by logical filters or on the basis of known data about the population of the area of interest. Further, the implementation of the TQA approach builds on the assumption that if boundary profiles are satisfied, all profiles within the studied segment of the population are covered.

To get as thorough of an insight as possible into a segment of the population in the context of this chapter, the daily routines of boundary profiles are important. There are as many routines as there are boundary profiles. However, there can be fewer different routines as there are profiles, as some profiles can have the same daily routines. An analysis of the daily routines of boundary profiles can result in the compatibility of various segments of the population in certain areas, as daily routines may explain similarities in people's interests. The implementation of TQA results in the acceptability and quality of places for a particular segment of the population, and it enables an examination of how well a certain place suits this group of people and how well it enables their co-habitation. The final result of the TQA approach is a time-quality balance of a profile.

Thus, the key points for any scenario or spatial development defined in this approach are the user profile, activity and space, where three conditions related to the user profile, acting in a certain territory, are analysed:

- time balance,
- economic balance and
- time-quality balance.

2.1. Time balance

Time balance shows how comfortably the user spends time in his/her (living) environment; how comfortably a segment of the population can live in a certain area, i.e. how a chosen routine is manageable for an individual in the available time frame; and whether a person can achieve necessary and optional activities within the available limits of the time frame (e.g. 24 h/day, within the schedules/opening hours of available capacities for the selected activities). A comparative analysis of more segments of the population shows the abilities of the co-habitation of various segments of the population in the same area. It also shows if any segment of the population is being disregarded or favoured. This means that the time-balance category is place dependent. This is closely linked with spatial characteristics (e.g. structure of the place, infrastructure, program, etc.). Time balance is possible to establish when one has a defined profile, routines and belonging space(s). Time balance shows how comfortably time is offered to the user through his/her (living) environment.

The time spent on each action should be shorter or equal to the available time for that action and should be accommodated within the time sequence available for the action:

$$T_{Rqi} \leq T_{Avi} \tag{1}$$

where T_{Rqi} = time required for action i; T_{Avi} = time available for action i.

For an illustration, when one does not manage that action in time, the person is late. However, the minimum required condition—although it is not always sufficient—is to do everything required in the entire available time frame (e.g. to do all daily routines in 24 h):

$$\sum_i T_{Rqi} \leq \sum_i T_{Avi} \rightarrow T_{Rq} \leq T_{Av} \tag{2}$$

Time-balance analysis shows a balance of necessary and optional activities. In the situation of assessing the suitability of a neighbourhood for a certain profile, first checking the criteria at the level of time balance is the profile's ability to fulfil activities. If the profile is not able to fulfil necessary activities, the neighbourhood is not suitable for it, and if the profile is not able to fulfil optional activities, optional activities must be re-organised against a new priority list.

2.2. Economic balance

Economic balance is a category that represents the subject's income and expenses for necessary and optional activities, and a financial framework within which the subject is flexible to be able to perform each of the activities in a certain environment, i.e. whether the selected activities can be afforded per person within a household and whether the incomes and expenses of a household per person enable these activities to be fulfilled.

The basic information addressed is a household's incomes and expenses for necessary activities and optional activities. The expenses of a household should not exceed the incomes:

$$\sum_i M_{Rqi} \leq \sum_j M_{Avj} \rightarrow M_{Rq} \leq M_{Av} \quad (3)$$

where M_{Rqi} = money required for expense i; M_{Avi} = money available from source j.

Incomes are classified as regular (e.g. salary earned in working time every working day); other regular (e.g. pension, rent); and irregular (e.g. property selling). Expenses are classified as: residential expenses; basic basket expenses (e.g. food, clothes); other necessary expenses (e.g. nursery, school); other optional expenses; and travel expenses for commuting as a part of a daily routine.

2.3. Time-quality balance

Time-quality balance calculates the time spent in terms of both activity and environment. The component of time-quality balance shows when a financial situation allows activities to happen; how well the time needed for them has been spent in total; and how much of the entire time taken for all of the activities per day is considered good quality and how much of it is bad quality. Time-quality balance shows the final quality of the time spent within a routine and reflects on the quality of the living environment in which the profile lives. Thus, with this final parameter, the TQA approach shows whether a segment of the population can live in a certain area and how comfortably:

$$K_{TQ} = \frac{T_Q}{T_{Sp}} = \frac{\sum_i T_{Qi}}{\sum_i T_{Spi}} = \frac{\sum_{ij} T_{Spi} \times F_{Qij} \times F_{Wij}}{\sum_i T_{Spi}} \quad (4)$$

where $\sum_j F_{Wij} = 1$ and $-1 \leq F_{Qij} \leq 1$; where K_{TQ} = time-quality coefficient; T_Q = evaluated portion of time (+ sign: good time; - sign: bad time); T_{Qi} = evaluated portion of time within time interval i; T_{Sp} = time spent; T_{Spi} = time spent within time interval i; F_{Qij} = quality of quality component j within time interval i; F_{Wij} = influence (weight) of quality component j within time interval i.

Following the TQA approach, two time-quality components are proposed:

AC = activity component; SC = space component

therefore

$$j \in \{AS, SC\} \Rightarrow F_{Wi,SC} = 1 - F_{Wi,AC} \quad (5)$$

2.4. Behaviour map: a means of TQA interpretation

To implement TQA as a universal evaluation tool for quality of place in relation to its usage, behaviour mapping [19] is seen as a key part of the process. This is true especially where behaviour maps extract behavioural evidence into layers of spatial information to provide a better understanding of the individual and the collective patterns of use that emerge in a place.

Behaviour mapping is a method and tool for analysing usage-spatial relationships originating from the field of environmental psychology in the 1970s of the twentieth century [20]. It is a product of observation and a tool for place analysis and design at the same time, and as such, it represents a means for recording behaviours in a spatial setting and the final results of the observation of dynamic patterns of spatial occupancies, visually expressing structural relations between the characteristics of places and their use(r)s. Behaviour maps can contain broad spectrums of information, from location, type of activity and duration of activity to many other characteristics, depending on the research question, aspects or issues. Therefore, they act as effective media for dealing with the spatial and dynamic patterns of the usage of places. Due to their graphic nature, visualising relationships between various—not necessarily physical—characteristics of places and their users, they can be seen as a valuable tool for improving bottom-up generated data and for providing new insights for spatial statistics. Practically, they can provide the recognition and understanding of possible or expected uses in places, their frequencies and their intensities, and as such, they may lead towards more effective and responsible planning and design practice and towards the achievement of better quality of living. Knowing actual activities in places and their characteristics in places is important for the identification of quality of everyday living and for directing and stimulating the suitability of territories for occupancy.

Some fundamental conditions need to be met before any recording of behaviour can start. It is necessary to define the area to be observed, to clearly define the types of activities and details about behaviours to be observed, to schedule specific times and their repetitions for observation, and to provide a system of recording, coding, counting and analysing with either a low- or high-tech recording approach. This chapter argues for behaviour mapping as both an analytical tool for monitoring daily routines and as a means of the interpretation of the TQA approach, and by this, it is promoted for the provision of bottom-up generated datasets as a basis for user-centred spatial statistics. Behaviour mapping has the capacity to address the social needs, locations, dimensions, frequency, intensity and co-habitation of activities in places directly. It refers to groups and individuals as well as social-relations change.

Thus, such behaviour maps can be used to capture the knowledge that brings the indirect insights of usage-spatial relationships and to visualise abstract notions and essentially the non-spatial characteristics of physical environments. In relation to TQA, one of the key pieces of information offered is time-related characteristics. A behaviour map can show two significant temporal dimensions: (1) for how long a certain activity is going on in a certain place and (2) on which day or in what other time-unit sequence the activity has been taking place. In the TQA approach, behaviour is usually defined by a daily routine but allows the consideration of other situations, e.g. a weekly routine and extraordinary routine.

The challenge of this concept is to shift the understanding about and focus on datasets for city analysis towards people and places. Assisted with behaviour mapping, this alternative approach provides a time-based perspective on the activities and engagement of people.

3. TQA implementation: case of Posavje District, Ljubljana, Slovenia

Posavje is one of Ljubljana's 17 districts; it features a wide range of spatial and living situations, from dense and high residential neighbourhoods to rural, mostly agricultural areas, and it includes a transport point of view supplied by public bus services and the regional railway track. To illustrate the TQA approach, the assessment of quality of living environments via quality of time for a family man was modelled. Time-quality assessment for the daily routine of a profile living in two different micro-locations within the same neighbourhood was analysed and simulated. Further, regarding the contemporary demographic situation across Europe, attention was paid to elderly people—one of the vulnerable user groups—assuming that when some settings and arrangements are good for them, it is quite likely that they may suit other users, too. Four different locations within four characteristic neighbourhoods in the area were analysed and simulated (**Figure 1**).

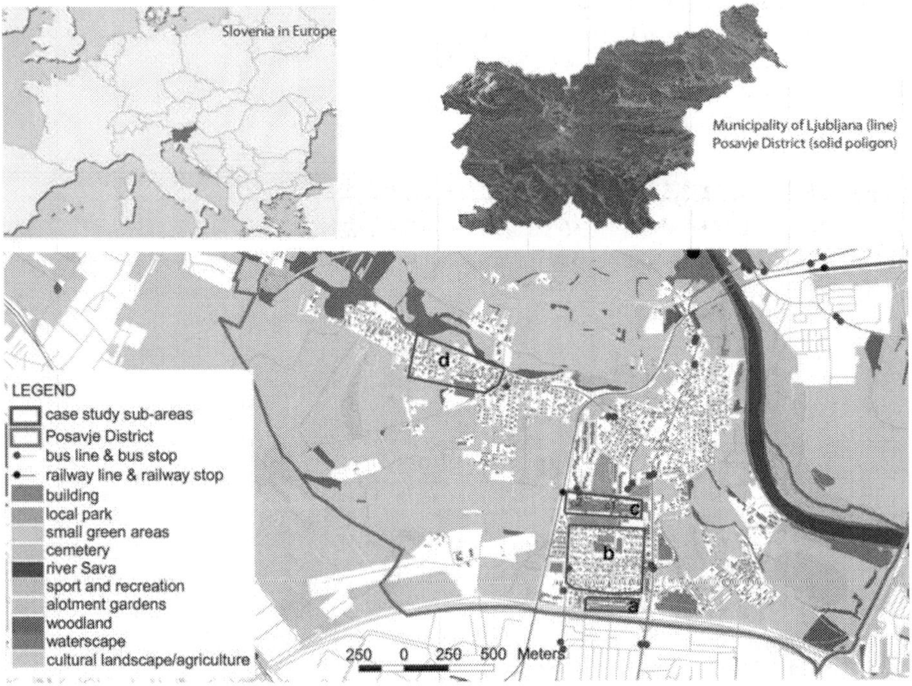

Figure 1. Case study area.

Time-quality assessment for the daily routine of a profile living in four different types of locations within the area was simulated using the TQA approach. High-rise flats that also provide accommodation unit for elderly people by the highway is denoted by the letter a; the area of individual houses by b; high-rise flats in the centre of the neighbourhood, close to the community centre, by c; and the area of a compact rural settlement by d, where d1 is assigned to the current state of d. The letter F denotes a profile of a working family man with pre-school children. The simulation examines two micro-locations in high-rise flats in the centre of the neighbourhood, close to the community centre, c1 and c2. The letter E denotes the profile of an elderly man; therefore, Ea, Eb, Ec1 and Ed1, respectively, are denotations of one of the possible daily routines of such a person, regarding the location of his home. Examples show daily routines in nice weather during spring or autumn.

The first case (Fc1 and Fc2; Section 3.1) is focused on the procedure of the TQA approach; setting up a profile; and defining and monitoring a routine and time-balance assessment, economic assessment and time-quality assessment as a final result of the process. Meanwhile, the second case (Ea, Eb, Ec1 and Ed1; Section 3.2) is focused on the characteristics of the routine of the profile living in different areas within the studied territory and their feasibility regarding the circumstances (Ed1–Ed4).

No absolute measure of quality of living space exists. One always compares two spaces to declare the quality of each, where one or both spaces may be fictive. The quality of one space may be defined in relation to another known or defined quality, whereas the parameters of quality depend on the purpose of the space and/or the user(s) of the space. Something that is important for one user may not be as important for another user or may not apply to other users at all.

The TQA approach extracts the time spent on any activity into the good or the bad portion. The rest of the time, not classified as good or bad, is considered as the indifferent portion of time. In the TQA approach, satisfaction with time is valuated using a scale from -100% satisfaction (complete dissatisfaction) to +100% satisfaction (complete satisfaction), where 0% satisfaction would mean that the user is indifferent to the time spent in a certain space.

To generalise in such a valuation:

- a good time is represented by a positively signed percentage of satisfaction;
- a bad time is represented by a negatively signed percentage of satisfaction; and
- the rest is identified as indifferent time.

This also indicates that a good time and a bad time can neutralize each other, e.g. good time = - bad time (e.g. 1 h of good time + 1 h of bad time = 0).

The measure of quality is the quantity of good time (or bad time if the result is negative) after summation. The sum of the absolute values of the quantity of time (good + bad + indifferent) may not exceed the absolute value of the available time (e.g. 24 h/day).

Further evaluation that introduces time as the universal measure for the quality of environments refers to activities and spaces linked to these two components of time (activity component—FQAC, spatial component—FQSC) by weight (FWAC, FWSC), i.e. how much of an influence each of the components has on the quality of time spent in a place for this certain activity. This thought depends on the profile's preferences, which may also depend on affordances (e.g. economic ability). The weight of each quality component describes how much each component contributes to the potential quality of time, e.g. the potential satisfaction with the time spent in the given place. These two parameters finally shape the activity-place relationship in a daily routine, and they are, for comparative purposes, finally transferred into the coefficient of time quality and time-quality balance (KTQ and TQ).

In all of the examples referring to the implementation of the TQA approach, the following parameters are assessed and/or calculated:

TSp, time spent (hours, minutes); FQAC, quality of activity component of time (%); FQSC, quality of spatial component of time (%); FWAC, influence of activity component of time (%); FWSC, influence of spatial component of time (%); KTQ, coefficient of time quality; TQ, quality time (hours, minutes).

When implementing the TQA approach, it must be remembered that time balance and economic balance are absolute objective measures, while time-quality balance is always subjective. Hence, it shows how one place may be better (e.g. provides higher benefit/comfort for the user) than the other and always needs to be commented regarding the context. In this respect, although economic balance represents an absolute value, it is linked to location.

3.1. Family man living in urban area

The simulation illustrates activity-place relations and time-quality balance (TQ) for a total daily routine for two variations of the same main socioeconomic profile from the same neighbourhood. The initial results are related to the time spent on the activities and the basic qualities of activities and places. Further evaluation introduces time as the measure for quality, referring to activities (activity component of time—FQAC) and places (spatial component of time—FQSC), taking into account the weight (FWAC, FWSC) of each quality component, which describes how much each component contributes to the potential quality of time. The final results are the coefficient of time quality (KTQ) and time-quality balance (TQ).

3.1.1. Profile and time balance

For an illustration, a segment of the population is presented. It is defined as an educated man with a permanent job and family. Age, family income and number of children are selected as three key characteristic parameters for setting up boundary profiles of such a segment of the population. The age ranged from the beginning of the carrier (30-year-old man) and towards the end of the carrier (55-year-old man). Boundaries regarding family incomes are represented by low-income educated family (2.400 EUR per month) and high-income educated family (12.000 EUR per month). Boundaries for the number of children are one child and four children. Based on these characteristics, eight combinations of profiles are possible (**Table 1**).

Discussing the possible daily routines of such eight profiles, generally, two different schedules exist: Those with more children spent more time on preparation activities and on dropping-off/picking-up activities. However, the assumption is that their final daily routines differ much more, including the time valuation of journeys among the activities, and the consideration of their working and opening hours.

3.1.2. Economic balance

The selected profile, used for an illustration of the TQA approach, is a member of a household characterised by: family with two adults (age 30–55), two children (age 1–15) and incomes (net) of 40.000 EUR/year, i.e. one parent earns 22.000 EUR, while the other earns 18.000 EUR,

	parameters of the profile						necessary stops at the daily routine							
	age		family income		children		home	home	nursery / school	work	shop	nursery / school	home	
profile	30 years	55 years	2400 per month	12000 per month	1	4	sleeping	preparation for go to work	dropping of children	working	daily shopping and services	picking up children	dinner preparation + dinner	time spent at all specific locations
E 1	x		x		x		8 h	30 min	5 min	8 h	20 min	5 min	30 min	**17h 30min**
E 2	x			x	x		8 h	30 min	5 min	8 h	20 min	5 min	30 min	**17h 30min**
E 3	x		x			x	8 h	45 min	10 min, 5 min	8 h	20 min	10 min, 5 min	45 min	**18h 20min**
E 4	x			x		x	8 h	45 min	10 min, 5 min	8 h	20 min	10 min, 5 min	45 min	**18h 20min**
E 5		x	x		x		8 h	30 min	5 min	8 h	20 min	5 min	30 min	**17h 30min**
E 6		x		x	x		8 h	30 min	5 min	8 h	20 min	5 min	30 min	**17h 30min**
E 7		x	x			x	8 h	45 min	10 min, 5 min	8 h	20 min	10 min, 5 min	45 min	**18h 20min**
E 8		x		x		x	8 h	45 min	10 min, 5 min	8 h	20 min	10 min, 5 min	45 min	**18h 20min**

Table 1. Parameters of the profile and its minimal time required for necessary activities in a daily routine, excluding commuting.

which equals approximately 11 EUR/working hour for the first and 9 EUR/working hour for the second.

Three characteristic situations are simulated (see **Table 2**). In the given neighbourhood, the basic level of expenses of such a household would be approximately 30.000 EUR/year. In the case where the family rents their apartment, their expenses are as follows (see case 1, **Table 2**): 12.000 EUR for residential expenses; 11.500 EUR for basic basket expenses (e.g. food, clothes); 4.400 EUR for other necessary expenses, such as nursery, school or the possession of a family car; 700 EUR for other optional expenses, such as hobbies, extra travel, vacations and extra insurances; and 1.400 EUR for commuting as a part of a daily routine, considering they are using public transport and they manage daily routines within the range of the city public transport area. In this case, the household may save 10.000 EUR/year = (40.000 – 30.000) EUR/year. However, their lives are rather ascetic.

In this case, if the family faced higher expenses (medium level), their earnings would soon become negligible or none. As simulated in **Table 2** (case 2), residential expenses are 15.600

	Earnings	Residential expenses	Basic basket	Other necessary expenses	Optional expenses	Commuting	Total	Savings
Case 1 [EUR]	40.000	12.000	11.500	4.400	700	1.400	30.000	10.000
Case 2 [EUR]	40.000	15.600	15.400	4.400	3.200	1.400	40.000	0
Case 3 [EUR]	40.000	4.600	15.400	4.400	3.200	1.400	29.000	11.000

Table 2. Examples of economic balance for three cases of the same profile.

EUR (12.000 EUR + 3.600 EUR), i.e. the family strives for better commodities and affords a larger apartment, assuring a room for every child. They increase the budget for basic basket goods to afford higher-quality products: 14.400 EUR (11.500 EUR + 3.900 EUR). For other necessary expenses, such as nursery, school or the possession of a family car, they spend the same as in case 1, 4.400 EUR. They put more of their budget towards other optional expenses, such as hobbies, extra travel, vacations and extra insurances, 3.200 EUR (700 EUR + 2.500 EUR), and they keep the same budget for travel expenses, 1.400 EUR. In this case, the balance is ±0 EUR/year = (40.000 – 40.000) EUR/year. This case (case 2, **Table 2**) illustrates the maximum standard that such a family could afford in the given neighbourhood. In the event they are satisfied with a less expensive apartment, they can accrue some savings. This can be achieved by changing the location or some other quality of the residence (e.g. size, building quality). However, this might increase time requirements for daily travel or decrease satisfaction during the time spent at home.

In the case the family owns the apartment (case 3, **Table 2**), the yearly residential expenses are considerably lower since the main expense (buying a flat) was realized in the past already. For a medium level of expenses, i.e. they spent 40.000 EUR/year, the savings would amount to approximately 11.000 EUR/year, allocating 1.000 EUR/year for the maintenance of their investment. In such a case, the considered family could easily afford a medium level of expenses or even a higher level (e.g. afford a better apartment or a second car). The question is the effect of each improvement on quality of living. The examples show that in the case where the incomes of such a household amount to less than 30.000 EUR/year and they do not have owner status but instead must rent an apartment, they could not afford to live in the given neighbourhood. In the case where they own an apartment, they could live there and even afford a slightly higher level of other expenses. Savings are usually also an important component of the financial security of a household and consequently influence satisfaction. Therefore, the ability of a household to create some savings in a given environment is not negligible.

3.1.3. Time-quality balance

Simulating time-quality balance for the same profile, with exactly the same daily routine, living in the same neighbourhood but on the other side, close to the railway line, would show that the time-quality balance would decrease. This is especially the case if the quality of the spatial component of time for sleeping, which represents a great portion of good quality time, is considered as rather bad. In such a case, instead of having 10 h 5′ (Fc1; $K_{TQ} = 0.42$) of a good quality of time per day, Fc2 has 8 h 38′ of a good quality of time per day ($K_{TQ} = 0.36$) (**Table 3**).

3.2. Elderly living in urban area

The profile was defined based on socioeconomic statistical data. The time and economic balance of the profile was assessed as positive. Data on time and activity were collected on the basis of a combination of approaches: field work related to spatial analysis, including facilities and services (e.g. open/green space, recreation, culture, public transport), and accessibility; and a pilot behaviour observation of the selected areas to get an idea of the behaviour patterns of elderly in the area, including the duration of activities in the environment (e.g. how

		Fc1									Fc2								
		T_{Sp}	T_{start}	T_{end}	F_{QAC}	F_{QSC}	F_{WAC}	F_{WSC}	K_{TQ}	T_Q	T_{Sp}	T_{start}	T_{end}	F_{QAC}	F_{QSC}	F_{WAC}	F_{WSC}	K_{TQ}	T_Q
home	sleeping	8 h 0 '	22:30	06:30	100	30	50	50	0,65	5 h 12 '	8 h 0 '	22:30	06:30	100	0	50	50	0,50	4 h 0 '
home	preparation for go to work	1 h 0 '	06:30	07:30	0	90	40	60	0,54	32 '	1 h 0 '	06:30	07:30	0	90	40	60	0,54	32 '
by bicycle	going to nursery	5 '	07:30	07:35	50	60	40	60	0,56	3 '	5 '	07:30	07:35	50	60	40	60	0,56	3 '
nursery / school	dropping of children	10 '	07:35	07:45	30	60	40	60	0,48	5 '	10 '	07:35	07:45	30	60	40	60	0,48	5 '
by bicycle	going to work	25 '	07:45	08:10	30	-10	40	60	0,06	2 '	25 '	07:45	08:10	30	-10	40	60	0,06	2 '
work	working	8 h 0 '	08:10	16:10	-20	0	50	50	-0,10	-48 '	8 h 0 '	08:10	16:10	-20	0	50	50	-0,10	-48 '
by bicycle	going from work, to pick up children	25 '	16:10	16:35	30	-10	40	60	0,06	2 '	25 '	16:10	16:35	30	-10	40	60	0,06	2 '
nursery / school	picking up children	15 '	16:35	16:50	50	60	40	60	0,56	8 '	15 '	16:35	16:50	50	60	40	60	0,56	8 '
by bicycle	going from work, to do shopping	5 '	16:50	16:55	50	60	40	60	0,56	3 '	5 '	16:50	16:55	50	60	40	60	0,56	3 '
shop	daily shopping and services	20 '	16:55	17:15	0	0	20	80	0,00	0 '	20 '	16:55	17:15	0	0	20	80	0,00	0 '
by bicycle	going home	5 '	17:15	17:20	30	60	40	60	0,48	2 '	5 '	17:15	17:20	30	60	40	60	0,48	2 '
by bicycle	going to recreation	5 '	17:20	17:25	70	60	50	50	0,65	3 '	5 '	17:20	17:25	70	60	50	50	0,65	3 '
gym	recreation + socialising	1 h 30 '	17:25	18:55	100	80	60	40	0,92	1 h 23 '	1 h 30 '	17:25	18:55	100	80	60	40	0,92	1 h 23 '
by bicycle	comming from recreation	5 '	18:55	19:00	70	60	50	50	0,65	3 '	5 '	18:55	19:00	70	60	50	50	0,65	3 '
home	dinner preparation + dinner	45 '	19:00	19:45	100	90	60	40	0,96	43 '	45 '	19:00	19:45	100	80	60	40	0,92	41 '
home	home with a family	2 h 45 '	19:45	22:30	100	80	60	40	0,92	2 h 32 '	2 h 45 '	19:45	22:30	100	60	60	40	0,84	2 h 19 '
		24 h 0 '							**0,42**	**10 h 5 '**	**24 h 0 '**							**0,36**	**8 h 38 '**

Table 3. Time quality balance for Fc1 and Fc2.

much time they spent coming from A to B, how much time they spent in a local park or library). An interview with an active member of the local community, an elderly person living in a high-raised flat area, was conducted and included questions about the daily routine there; the environmental, social and economic commodities associated with living there; and the like.

Parameter calibration was done with a combination of discipline tacit knowledge, expert knowledge and target-group involvement; the space component as a combination of field work, cartographic materials, expert knowledge and target-group involvement (indirectly with behaviour mapping, directly via interviewing); and the activity component of target-group involvement (indirectly with behaviour mapping, directly via interviewing).

		home	home	public space O	cafe	public space O	shop	public space O	home	home	home	public space O	public space O	public space	bus	public space O	public space I	public space O	bus	public space O	home	home	
		sleeping	morning at home + breakfast	walking to the cafe	socialising	walking to the grocery store	shopping	walking back home	cooking	having lunch	resting at home	walking in the neighbourhoo d	relaxing in the park / green area	walking to the bus station	travelling by bus	walking to the library	in the library	walking from the library	travelling by bus	walking home	dinner preparation + dinner	easy evening at home	
Ea	T_{Sp}	8 h 0'	2 h 30'	10'	1 h 30'	0'	30'	10'	1 h 30'	1 h 0'	1 h 30'	15'	1 h 35'	0'	0'	10'	1 h 0'	5'	30'	5'	1 h 0'	2 h 30'	**24 h 0'**
	T_{start}	22:30	06:30	09:00	09:10	10:40	10:40	11:10	11:20	12:50	13:50	15:20	15:35	17:10	17:10	17:10	17:20	18:20	18:25	18:55	19:00	20:00	
	T_{end}	06:30	09:00	09:10	10:40	10:40	11:10	11:20	12:50	13:50	15:20	15:35	17:10	17:10	17:10	17:20	18:20	18:25	18:55	19:00	20:00	22:30	
	F_{QAC}	100	80	60	80	40	25	30	60	80	85	75	90	60	-10	60	100	40	-10	40	80	100	
	F_{QSC}	-20	60	20	65	20	0	20	60	60	-20	30	80	60	-10	60	80	50	-10	10	60	40	
	F_{WAC}	50	50	50	60	50	60	50	50	50	40	40	40	50	70	50	40	50	70	50	50	40	
	F_{WSC}	50	50	50	40	50	40	50	50	50	60	60	60	50	30	50	60	50	30	50	50	60	
	K_{TQ}	0,40	0,70	0,40	0,74	0,30	0,15	0,25	0,60	0,70	0,22	0,48	0,84	0,60	-0,10	0,60	0,88	0,45	-0,10	0,25	0,70	0,64	**0,54**
	T_Q	3 h 12'	1 h 45'	4'	1 h 7'	0'	5'	3'	54'	42'	20'	7'	1 h 20'	0'	0'	6'	53'	2'	-3'	1'	42'	1 h 36'	**12 h 55'**
Eb	T_{Sp}	8 h 0'	2 h 30'	5'	1 h 30'	0'	30'	5'	1 h 30'	1 h 0'	1 h 30'	5'	2 h 5'	0'	0'	10'	1 h 0'	5'	20'	5'	1 h 0'	2 h 30'	**24 h 0'**
	T_{start}	22:30	06:30	09:00	09:05	10:35	10:35	11:05	11:10	12:40	13:40	15:10	15:15	17:20	17:20	17:20	17:30	18:30	18:35	18:55	19:00	20:00	
	T_{end}	06:30	09:00	09:05	10:35	10:35	11:05	11:10	12:40	13:40	15:10	15:15	17:20	17:20	17:20	17:30	18:30	18:35	18:55	19:00	20:00	22:30	
	F_{QAC}	100	80	60	80	40	25	30	60	80	85	75	90	60	-10	60	100	40	-10	40	80	100	
	F_{QSC}	50	90	25	65	25	0	25	90	90	50	35	80	60	0	60	80	50	0	10	90	90	
	F_{WAC}	50	50	50	60	50	60	50	50	50	40	40	40	50	70	50	40	50	70	50	50	40	
	F_{WSC}	50	50	50	40	50	40	50	50	50	60	60	60	50	30	50	60	50	30	50	50	60	
	K_{TQ}	0,75	0,85	0,43	0,74	0,33	0,15	0,28	0,75	0,85	0,64	0,51	0,84	0,60	-0,07	0,60	0,88	0,45	-0,07	0,25	0,85	0,94	**0,76**
	T_Q	6 h 0'	2 h 8'	2'	1 h 7'	0'	5'	1'	1 h 8'	51'	58'	3'	1 h 45'	0'	0'	6'	53'	2'	-1'	1'	51'	2 h 21'	**18 h 19'**

Table 4. Time quality balance for Ea and Eb.

http://dx.doi.org/10.5772/65322

3.2.1. Time-quality balance analysis for profile from various locations

The results in **Tables 4** and **5** indicate that the best living conditions for an elderly person are the areas of b and c, while area a is disadvantageous primarily due to highway pollution (noise, air pollution) and partly due to remoteness regarding the community/neighbourhood centre. Area d is a bit remote, which is significant for agricultural production activities (early morning noise, season noise, smell), a mixed zone of living-agriculture-small-industry

Ec1

Location	Activity	T_{Sp}	T_{start}	T_{end}	F_{QAC}	F_{QSC}	F_{WAC}	F_{WSC}	K_{TQ}	T_Q
home	sleeping	8 h 0'	22:30	06:30	100	30	50	50	0,65	5 h 12'
home	morning at home + breakfast	2 h 30'	06:30	09:00	80	90	50	50	0,85	2 h 8'
public space O	walking to the cafe	5'	09:00	09:05	60	60	50	50	0,60	3'
cafe	socialising	1 h 30'	09:05	10:35	80	65	60	40	0,74	1 h 7'
public space O	walking to the grocery store	2'	10:35	10:37	40	60	50	50	0,50	1'
shop	shopping	30'	10:37	11:07	25	0	60	40	0,15	5'
public space O	walking back home	5'	11:07	11:12	30	60	50	50	0,45	2'
home	cooking	1 h 30'	11:12	12:42	60	90	50	50	0,75	1 h 8'
home	having lunch	1 h 0'	12:42	13:42	80	90	50	50	0,85	51'
home	resting at home	1 h 30'	13:42	15:12	85	30	40	60	0,52	47'
public space O	walking in the neighbourhood	5'	15:12	15:17	75	60	40	60	0,66	3'
public space O	relaxing in the park / green area	2 h 3'	15:17	17:20	90	80	40	60	0,84	1 h 43'
public space	walking to the bus station	0'	17:20	17:20	60	60	50	50	0,60	0'
bus	travelling by bus	0'	17:20	17:20	-10	0	70	30	-0,07	0'
public space O	walking to the library	10'	17:20	17:30	60	60	50	50	0,60	6'
public space I	in the library	1 h 22'	17:30	18:52	100	80	40	60	0,88	1 h 12'
public space O	walking from the library	0'	18:52	18:52	50	60	50	50	0,55	0'
bus	travelling by bus	0'	18:52	18:52	-10	0	70	30	-0,07	0'
public space O	walking home	8'	18:52	19:00	50	60	50	50	0,55	4'
home	dinner preparation + dinner	1 h 0'	19:00	20:00	80	90	50	50	0,85	51'
home	easy evening at home	2 h 30'	20:00	22:30	100	80	40	60	0,88	2 h 12'
		24 h 0'							**0,73**	**17 h 34'**

Ed1

Location	Activity	T_{Sp}	T_{start}	T_{end}	F_{QAC}	F_{QSC}	F_{WAC}	F_{WSC}	K_{TQ}	T_Q
home	sleeping	8 h 0'	22:30	06:30	100	20	50	50	0,60	4 h 48'
home	morning at home + breakfast	2 h 30'	06:30	09:00	80	80	50	50	0,80	2 h 0'
public space O	walking to the cafe	5'	09:00	09:05	60	-10	50	50	0,25	1'
cafe	socialising	1 h 30'	09:05	10:35	80	80	60	40	0,80	1 h 12'
public space O	walking to the grocery store	15'	10:35	10:50	40	-10	50	50	0,15	2'
shop	shopping	30'	10:50	11:20	35	0	60	40	0,21	6'
public space O	walking back home	15'	11:20	11:35	40	-10	50	50	0,15	2'
home	cooking	1 h 30'	11:35	13:05	60	80	50	50	0,70	1 h 3'
home	having lunch	1 h 0'	13:05	14:05	80	80	50	50	0,80	48'
home	resting at home	1 h 30'	14:05	15:35	85	20	40	60	0,46	41'
public space O	walking in the neighbourhood	5'	15:35	15:40	75	30	40	60	0,48	2'
public space O	relaxing in the park / green area	1 h 30'	15:40	17:10	75	60	40	60	0,66	59'
public space	walking to the bus station	0'	17:10	17:10	60	30	50	50	0,45	0'
bus	travelling by bus	10'	17:10	17:20	-10	0	70	30	-0,07	-1'
public space O	walking to the library	5'	17:20	17:25	40	50	50	50	0,45	2'
public space I	in the library	1 h 10'	17:25	18:35	100	80	40	60	0,88	1 h 2'
public space O	walking from the library	5'	18:35	18:40	40	50	50	50	0,45	2'
bus	travelling by bus	10'	18:40	18:50	-10	0	70	30	-0,07	-1'
public space O	walking home	10'	18:50	19:00	40	-10	50	50	0,15	2'
home	dinner preparation + dinner	1 h 0'	19:00	20:00	80	80	50	50	0,80	48'
home	easy evening at home	2 h 30'	20:00	22:30	100	60	40	60	0,76	1 h 54'
		24 h 0'							**0,65**	**15 h 34'**

Table 5. Time quality balance for Ec1 and Ed1.

Activity	F_{WAC}	F_{WSC}	Ed2 T_{Sp}	Ed2 T_{start}	Ed2 T_{end}	Ed2 F_{QAC}	Ed2 F_{QSC}	Ed2 K_{TQ}	Ed2 T_Q
sleeping	50	50	8 h 0 '	22:30	06:30	100	20	0,60	4 h 48 '
morning at home + breakfast	50	50	2 h 30 '	06:30	09:00	80	80	0,80	2 h 0 '
walking to the cafe	50	50	5 '	09:00	09:05	60	-10	0,25	1 '
socialising	60	40	1 h 30 '	09:05	10:35	80	80	0,80	1 h 12 '
walking to the grocery store	50	50	15 '	10:35	10:50	40	-10	0,15	2 '
shopping	60	40	30 '	10:50	11:20	35	0	0,21	6 '
walking back home	50	50	15 '	11:20	11:35	40	-10	0,15	2 '
cooking	50	50	1 h 30 '	11:35	13:05	60	80	0,70	1 h 3 '
having lunch	50	50	1 h 0 '	13:05	14:05	80	80	0,80	48 '
resting at home	40	60	1 h 30 '	14:05	15:35	85	20	0,46	41 '
walking in the neighbourhood	40	60	5 '	15:35	15:40	75	30	0,48	2 '
relaxing in the park / green area	40	60	1 h 15 '	15:40	16:55	75	60	0,66	50 '
walking to the bus station	50	50	0 '	16:55	16:55	60	30	0,45	0 '
travelling by bus	70	30	0 '	16:55	16:55	-10	0	-0,07	0 '
walking to the library	50	50	20 '	16:55	17:15	20	-20	0,00	0 '
in the library	40	60	1 h 10 '	17:15	18:25	100	80	0,88	1 h 2 '
walking from the library	50	50	0 '	18:25	18:25	40	50	0,45	0 '
travelling by bus	70	30	0 '	18:25	18:25	-10	0	-0,07	0 '
walking home	50	50	35 '	18:25	19:00	-10	-30	-0,20	-7 '
dinner preparation + dinner	50	50	1 h 0 '	19:00	20:00	80	80	0,80	48 '
easy evening at home	40	60	2 h 30 '	20:00	22:30	100	60	0,76	1 h 54 '
			24 h 0 '					**0,63**	**15 h 13 '**

Activity	Ed3 T_{Sp}	Ed3 T_{start}	Ed3 T_{end}	Ed3 F_{QAC}	Ed3 F_{QSC}	Ed3 K_{TQ}	Ed3 T_Q
sleeping	8 h 0 '	22:30	06:30	100	20	0,60	4 h 48 '
morning at home + breakfast	2 h 30 '	06:30	09:00	80	80	0,80	2 h 0 '
walking to the cafe	5 '	09:00	09:05	60	-10	0,25	1 '
socialising	1 h 0 '	09:05	10:05	80	80	0,80	48 '
walking to the grocery store	30 '	10:05	10:35	-10	0	-0,05	-2 '
shopping	30 '	10:35	11:05	25	0	0,15	5 '
walking back home	35 '	11:05	11:40	-30	-5	-0,18	-6 '
cooking	1 h 30 '	11:40	13:10	60	80	0,70	1 h 3 '
having lunch	1 h 0 '	13:10	14:10	80	80	0,80	48 '
resting at home	1 h 40 '	14:10	15:50	85	20	0,46	46 '
walking in the neighbourhood	5 '	15:50	15:55	75	30	0,48	2 '
relaxing in the park / green area	1 h 15 '	15:55	17:10	75	60	0,66	50 '
walking to the bus station	0 '	17:10	17:10	60	30	0,45	0 '
travelling by bus	10 '	17:10	17:20	-10	0	-0,07	-1 '
walking to the library	5 '	17:20	17:25	40	50	0,45	2 '
in the library	1 h 10 '	17:25	18:35	100	80	0,88	1 h 2 '
walking from the library	5 '	18:35	18:40	40	50	0,45	2 '
travelling by bus	10 '	18:40	18:50	-10	0	-0,07	-1 '
walking home	10 '	18:50	19:00	40	-10	0,15	2 '
dinner preparation + dinner	1 h 0 '	19:00	20:00	80	80	0,80	48 '
easy evening at home	2 h 30 '	20:00	22:30	100	60	0,76	1 h 54 '
	24 h 0 '					**0,62**	**14 h 51 '**

Activity	Ed4 T_{Sp}	Ed4 T_{start}	Ed4 T_{end}	Ed4 F_{QAC}	Ed4 F_{QSC}	Ed4 K_{TQ}	Ed4 T_Q
sleeping	8 h 0 '	22:30	06:30	100	20	0,60	4 h 48 '
morning at home + breakfast	2 h 30 '	06:30	09:00	80	80	0,80	2 h 0 '
walking to the cafe	5 '	09:00	09:05	60	-10	0,25	1 '
socialising	1 h 0 '	09:05	10:05	80	80	0,80	48 '
walking to the grocery store	30 '	10:05	10:35	-10	-10	-0,10	-3 '
shopping	30 '	10:35	11:05	25	0	0,15	5 '
walking back home	40 '	11:05	11:45	-30	-10	-0,20	-8 '
cooking	1 h 30 '	11:45	13:15	60	80	0,70	1 h 3 '
having lunch	1 h 0 '	13:15	14:15	80	80	0,80	48 '
resting at home	2 h 0 '	14:15	16:15	85	20	0,46	55 '
walking in the neighbourhood	5 '	16:15	16:20	75	30	0,48	2 '
relaxing in the park / green area	35 '	16:20	16:55	75	60	0,66	23 '
walking to the bus station	0 '	16:55	16:55	60	30	0,45	0 '
travelling by bus	0 '	16:55	16:55	-10	0	-0,07	0 '
walking to the library	20 '	16:55	17:15	20	-20	0,00	0 '
in the library	1 h 10 '	17:15	18:25	100	80	0,88	1 h 2 '
walking from the library	0 '	18:25	18:25	40	50	0,45	0 '
travelling by bus	0 '	18:25	18:25	-10	0	-0,07	0 '
walking home	35 '	18:25	19:00	-10	-30	-0,20	-7 '
dinner preparation + dinner	1 h 0 '	19:00	20:00	80	80	0,80	48 '
easy evening at home	2 h 30 '	20:00	22:30	100	60	0,76	1 h 54 '
	24 h 0 '					**0,60**	**14 h 19 '**

Table 6. Time quality balance for Ed2, Ed3 and Ed4.

activities, relative weakness in the supply of daily goods, poor capacity and poor management of the spatial infrastructure and therefore also traffic safety issues. However, in comparison to area a, the major advantages are direct contact with green areas, slightly better logistics towards the library and local community centre and lower traffic influence.

3.2.2. Time-quality balance simulation for profile from rural area in the case of changes

Ed2, Ed3 and Ed4 are simulations of daily routines of the profile in the case of the degradation of area d (**Table 6**). Ed2 simulates a situation where the end bus stop is cancelled, so the area is no longer provided with public transport. Ed3 simulates a situation in which the local supply of daily goods (which is of poor quality already) is completely cancelled, whereas Ed4 simulates a situation in which the area is without a bus and a local grocery supply.

Simulated changes indicate a similar decrease of the comfort of the feasibility of the examined routine due to the cancellation of either a bus (Ed2) or a local grocery supply (Ed3). In the case of the cancelation of both facilities, the daily schedule has to be modified, which manifests in time balance (e.g. less socialising and afternoon green-area walking, more necessary walking [commuting] and resting). In this simulation, the profile finally loses 1 h 15′ quality time. However, in the Ed4 situation, the routine, which includes shopping and library visiting, is feasible only in good weather conditions, while in the cases of Ea, Eb, Ec1 and Ed1, such a routine is also manageable in other weather circumstances.

4. Discussion

Implementing the TQA approach results in several levels of outputs, i.e. several evidence-based user-centred data available to inform the spatial statistics of territories. They are data on time balance, data on economic balance and data on time-quality balance.

Such data are linked to both locations and profiles. They enable one to compare profiles within different locations in the area or to inform about the suitability of a certain location in the area for various profiles. Further, they indicate a comparative suitability level of a location for living for a chosen profile against some other location for the same profile, as well as the suitability of a location for one profile in comparison to another.

In providing sufficient repeated analyses or simulations (taking into account various circumstances and edge conditions; e.g. weather conditions), such results can be visualised on a behaviour map showing a profile's suitability map for living. When more profiles are involved, a suitability map for living of a community with certain characteristics (minimum profile—the weakest link; average profile—general public in the area) in an area is the final output. Moreover, results can also show which profile can reach the minimum satisfaction level at a certain location in the area and the mapping suitability for the weakest profiles of the community, where different profiles are recognised as the weakest at different locations within the studied area.

Information offered by the TQA approach is useful for any kind of place user, from individuals to check locations, e.g. where to live or work, to decision-makers at various governance levels. The distribution of such information is possible through the upgrade of existing available information systems. Such information is under a constant refinement process referring to two main sources: available geoinformatics and spatial data, and direct and indirect participatory data. TQA as a monitoring or development-control approach is applicable to authorities and individuals for establishing new developments in a place, searching for measures of improvements, the comparison of different locations for one particular use and the comparison of various measures in a certain location.

5. Conclusion

This chapter presents and debates a spatial interaction approach for collecting, analysing and monitoring evidence-based data to assess quality of space for a certain use (activity) and a certain user (profile) via analysis of the quality of time spent on that activity in a particular space or sequence of spaces, using the TQA approach. The TQA approach proposes time as the universal expression and measure of quality of living, using time balance, economic balance and time-quality balance as the key indicators for calculating the possibility and comfort of living in the given environment. Data as a result of such an approach are linked to locations and user profiles and are therefore useful for the comparison of profiles within different locations of the area, and judgement about the suitability of certain locations in the area for various profiles.

It illustrates activity-place relations and time-quality balance (TQ) for the total daily routine for variations of the same main socioeconomic profile from the same neighbourhood. The initial results are related to the time spent on the activities and the basic qualities of activities and places. Further evaluation introduces time as the measure for quality, referring to activities (activity component of time—FQAC) and places (spatial component of time—FQSC), taking into account the weight (FWAC, FWSC) of each quality component, which describes how much each component contributes to the potential quality of time. The final results are the coefficient of time quality (KTQ) and time-quality balance (TQ).

The applicable value of this approach is in showing the suitability of a certain location for a chosen profile in comparison with some other location for the same profile, or in showing the suitability of a location for one profile in comparison with another. This is especially important in spatial planning and investment decision-making processes, as simulating a community with certain characteristics represented via a bunch of profiles (e.g., minimum profile—the weakest link; average profile—general public in the area) allows for a comprehensive simulation of living conditions for future residents or other (business) users. In this respect, the TQA approach can be used for searching for measures for improvements in territories, the comparison of different locations for one particular use, the comparison of various measures in a certain location and establishing new developments in a place. The capability of contemporary ICT tools that serve as an interface between place and people can play a significant role in automating data. Especially, monitoring tools consisting of a smartphone application, a set of

Web services and cloud computing and storage can provide very informative and rich information about the parameters relevant for the TQA approach. Such technology (e.g. [21]) enables insights into a real bottom-up understanding of the daily routines and circumstances with which people are involved, and it is worth linking with TQA in the further development of the approach and its implementation.

Author details

Damjan Marušić[1] and Barbara Goličnik Marušić[2]*

*Address all correspondence to: barbara.golicnik-marusic@uirs.si

1 DIPSTOR Ltd., Ljubljana, Slovenia

2 Urban Planning Institute of the Republic of Slovenia, Ljubljana, Slovenia

References

[1] Costanza R, Wainger L, Folke C, Maler K G. Modelling complex ecological economic systems: Toward an evolutionary, dynamic understanding of people and nature. BioScience. 1993;43:545–555.

[2] Comfort L K, Sungu Y, Johnson D, Dunn M. Complex systems in crisis: Anticipation and resilience in dynamic environments. Journal of Contingencies and Crisis Management. 2001;9(3):144–158.

[3] Grimm V, Revilla E, Berger U, Jeltsch F, Mooij F W, Railsback S F, Thulke H H, Weiner J, Wiegand T, DeAngelis D L. Pattern-oriented modeling of agent-based complex systems: lessons from ecology. Science. 2005;310(5750):987–991.

[4] Bar-Yam Y. Complexity Rising: From Human Beings to Human Civilization. A Complexity Profile, Encyclopedia of Life Support Systems. Oxford, UK: EOLSSUNESCO Publishers; 2002.

[5] Encyclopedia Britannica. Complexity [Internet]. 2016. Available from: https://www.britannica.com/science/complexity-scientific-theory [Accessed: 2016-07-04]

[6] Polajnar M, Demšar J. Small network completion using frequent subnetworks. Intelligent Data Analysis. 2015;19(1):89–108.

[7] Chua A, Marcheggiani E, Servillo L, Vande Moere A. FlowSampler: Visual analysis of urban flows in geolocated social media data. In: Maria Aiello L, McFarland D, editors. International Conference on Social Informatics (SocInfo'14); 10–13 November 2014; Barcelona: Springer; 2014. pp. 5–17.

[8] Pucci P, Manfredini F, Tagliolato P, editors. Mapping Urban Practices through Mobile Phone Data. Springer Briefs in Applied Sciences and Technology, Politecnico di Milano: Springer; 2015. 90 p.

[9] Dijst M, Vidakovic V. Travel ratio time: The key factor of spatial research. Transportation. 2000;27(2):179–199.

[10] Allen L R, Gibson R. Perceptions of community life and services: a comparison between leaders and community residents. Journal of the Community Development Society. 1987;18(1):89–103.

[11] Norris T. America's community movement: Investing in the civic landscape. American Journal of Community Psychology. 2001;29(2):301–307.

[12] Oort F. Using structural equation modelling to detect response shifts and true change. Quality of Life Research. 2005;14(3):587–598.

[13] Baker D A, Palmer R J. Examining the effects of perceptions of community and recreation participation on quality of life. Social Indicators Research. 2006;75:396–418.

[14] Blomquist G C. Measuring quality of life. In: Arnott R J, McMillen D P, editors. A Companion of Urban Economics. London: Blackwell Publishing; 2006. pp. 483–501.

[15] Lora E, Powell A. A new way of monitoring the quality of urban life. UNU-WIDER Working Paper. 2011;12:1–24.

[16] The Organisation for Economic Co-operation and Development (OECD). Guidelines on Measuring Subjective Well-being [Internet]. 2013. Available from: http://dx.doi.org/10.1787/9789264191655-en [Accessed 2014-03-23]

[17] Leipzig Charter. Leipzig Charter on Sustainable European Cities. Leipzig; 2007.

[18] Marušić D, Goličnik Marušić, B. Model for valuation and simulation of quality of living environments. International Journal of Innovation and Regional Development. 2014;5(4/5):405–428.

[19] Goličnik Marušić B, Marušić D. Behavioural maps and GIS in place evaluation and design. In: Alam B M, editor. Application of Geographic Information Systems. Rijeka: InTech; 2012. pp. 113–139.

[20] Ittelson W H, Rivlin L G, Prohansky H M. The use of behavioural maps in environmental psychology. In: Prohansky H M, Ittelson W H, Rivlin L G, editors. Environmental Psychology: Man and His Physical Setting. New York: Holt, Rinehart & Winston; 1970. pp. 658–668.

[21] Starič A, Demšar J, Zupan, B. Concurrent software architectures for exploratory data analysis. Wiley Interdisciplinary Reviews, Data Mining and Knowledge Discovery. 2015;5(4):165–180.